George John Romanes

Jelly-Fish, Star-Fish, and Sea-Urchins

Being a research on primitive nervous systems

George John Romanes

Jelly-Fish, Star-Fish, and Sea-Urchins
Being a research on primitive nervous systems

ISBN/EAN: 9783337037307

Printed in Europe, USA, Canada, Australia, Japan

Cover: Foto ©berggeist007 / pixelio.de

More available books at **www.hansebooks.com**

THE INTERNATIONAL SCIENTIFIC SERIES.

JELLY-FISH, STAR-FISH AND SEA-URCHINS

BEING

A RESEARCH ON PRIMITIVE NERVOUS SYSTEMS

BY

G. J. ROMANES, M. A., LL. D., F. R. S.

ZOÖLOGICAL SECRETARY OF THE LINNEAN SOCIETY

NEW YORK:
D. APPLETON AND COMPANY,
1, 3, AND 5 BOND STREET.
1893.

PREFACE.

When I first accepted the invitation of the editors of the International Scientific Series to supply a book upon Primitive Nervous Systems, I intended to have supplemented the description of my own work on the physiology of the *Medusæ* and *Echinodermata* with a tolerably full exposition of the results which have been obtained by other inquirers concerning the morphology and development of these animals. But it soon became apparent that it would be impossible, within the limits assigned to me, to do justice to the more important investigations upon these matters; and therefore I eventually decided upon restricting this essay to an account of my own researches.

With the exception of a few woodcuts in the last chapter (for the loan of which I am indebted to the kindness of Messrs. Cassell), all the illustrations are either original or copies of those in my Royal Society papers. In the letter-press also I have not scrupled to draw upon these papers, wherever it seemed to me that the passages would be sufficiently intelligible to a general reader. I may observe, however, that although I have throughout kept in view the requirements of a general reader, I have also sought to render the book of service to the working physiologist, by bringing together in one consecutive account all the more important observations and results which have been yielded by this research.

G. J. R.

London, 1884.

CONTENTS.

CHAPTER		PAGE
	INTRODUCTION	1
I.	STRUCTURE OF THE MEDUSÆ	10
II.	FUNDAMENTAL EXPERIMENTS	26
III.	EXPERIMENTS IN STIMULATION	37
IV.	EXPERIMENTS IN SECTION OF COVERED-EYED MEDUSÆ	65
V.	EXPERIMENTS IN SECTION OF NAKED-EYED MEDUSÆ	104
VI.	CO-ORDINATION	130
VII.	NATURAL RHYTHM	146
VIII.	ARTIFICIAL RHYTHM	175
IX.	POISONS	213
X.	STAR-FISH AND SEA-URCHINS	251

JELLY-FISH, STAR-FISH, AND SEA-URCHINS.

INTRODUCTION.

AMONG the most beautiful, as well as the most common, of the marine animals which are to be met with upon our coasts are the jelly-fish and the star-fish. Scarcely any one is so devoid of the instincts either of the artist or of the naturalist as not to have watched these animals with blended emotions of the æsthetic and the scientific—feeling the beauty while wondering at the organization. How many of us who live for most of the year in the fog and dust of large towns enjoy with the greater zest our summer's holiday at the seaside? And in the memories of most of us is there not associated with the picture of breaking waves and sea-birds floating indifferently in the blue sky or on the water still more blue, the thoughts of many a ramble among the weedy rocks and living pools, where for the time being we all become naturalists, and where those who least know what they are likely to find in their search are most likely to approach the keen happiness of childhood? If so, the image of the red sea-stars bespangling a mile of shining sand, or decorating the darkness of a thousand grottoes, must be joined with the image, no less vivid, of those crystal globes pulsating with life and gleaming with all the colours of the rainbow, which are perhaps the most strange, and certainly in my estimation the most delicately lovely creatures in the world.

It is with these two kinds of creatures that the present work is concerned, and if it seems almost impious to lay the "forced fingers rude" of science upon living things of such exquisite beauty, let it be remembered that our human nature is not so much out of joint that the rational desire to know is incompatible with the emotional impulse to admire. Speaking for myself, I can testify that my admiration of the extreme beauty of these animals has been greatly enhanced—or rather I should say that this extreme beauty has been, so to speak, revealed—by the continuous and close observation which many of my experiments required: both with the unassisted eye and with the microscope numberless points of detail, unnoticed before, became familiar to the mind; the forms as a whole were impressed upon the memory; and, by constantly watching their movements and changes of appearance, I have grown, like an artist studying a face or a landscape, to appreciate a fulness of beauty, the *esse* of which is only rendered possible by the *per cipi* of such attention as is demanded by scientific research. Moreover, association, if not the sole creator, is at least a most important factor of the beautiful; and therefore the sight of one of these animals is now much more to me, in the respects which we are considering, than it can be to any one in whose memory it is not connected with many days of that purest form of enjoyment which can only be experienced in the pursuit of science.

And here I may observe that the worker in marine zoology has one great advantage over his other scientific brethren. Apart from the intrinsic beauty of most of the creatures with which he has to deal, all the accompaniments of his work are æsthetic, and removed from those more or less offensive features which are so often necessarily incidental to the study of anatomy and physiology in the higher animals. When, for instance, I contrast my own work in a town laboratory on vertebrated animals with that which I am now about to describe upon the invertebrated in a laboratory set up upon the sea-beach, it is impossible not to feel that the contrast in point of enjoyment is considerable. In the latter case, a summer's work resembles the pleasure-making of a picnic prolonged for months, with the sense of feeling all the while that no time is being profitlessly spent. Whether one is sailing about upon the sunny sea, fishing with muslin nets for the surface fauna, or steaming away far from shore to dredge for other material, or, again, carrying on observations in the cool sea-water tanks and bell-jars of a neat little wooden workshop thrown open to the sea-breezes, it alike requires some effort to persuade one's self that the occupation is really something more than that of finding amusement.

It is now twelve years since I first took to this kind of summer recreation, and during that time most of my attention while at the seaside has been devoted to the two classes of animals already mentioned—viz. the jelly-fish and star-fish, or, as naturalists have named them, the Medusæ and Echinodermata. The present volume contains a tolerably full account of the results which during six of these summers I have succeeded in obtaining. If any of my readers should think that the harvest appears to be a small one in relation to the time and labour spent in gathering it, I shall feel pretty confident that those readers are not themselves working physiologists, and, therefore, that they are really ignorant of the time and labour required to devise and execute even apparently simple experiments, to hunt down a physiological question to its only possible answer, and to verify each step in the process of an experimental proof. Moreover, the difficulties in all these respects are increased tenfold in a seaside laboratory without adequate equipments or attendance, and where, in consequence, more time is usually lost in devising make-shifts for apparatus, and teaching unskilled hands how to help, than is consumed in all other parts of a research. From the picnic point of view, however, there is no real loss in this; such incidental difficulties add to the enjoyment (else why choose to make an extemporized grate and boil a kettle in the wood, when a much more efficient grate, full of lighted coals, is already boiling some other kettle at home?); and if they somewhat unduly prolong a research, the full meaning of life is, after all, not exhausted by the experiences of a mill-horse, and it is well to remember that so soon as we cease to take pleasure in our work, we are most likely sacrificing one part of our humanity to the altar of some other, and probably less worthy, constituent.

I may now say a few words on the scope of the investigations which are to be described in the present treatise. To some extent this is conveyed by the title; but I may observe that, as the "primitive nervous systems" whose physiology I have sought to advance are mainly subservient to the office of locomotion, in my Royal Society papers upon these researches I have adopted the title of "Observations on the Locomotor System" of each of the classes of animals in question. It is of interest to notice in this connection that the plan or mechanism of locomotion is completely different in the two classes, and that in the case of each class the plan or mechanism is unique, *i.e.* is not to be met with elsewhere in the animal kingdom. It is curious, however, that, in the case of one family of star-fish (the *Comatulæ*), owing to an extreme modification of form and function presented by the constituent parts of the locomotor organs, the method of progression has come closely to resemble that which is characteristic of jelly-fish.

There is still one preliminary topic on which I

feel that it is desirable to touch before proceeding to give an account of my experiments, and this has reference to the vivisection which many of these experiments have entailed. But in saying what I have to say in this connection I can afford to be brief, inasmuch as it is not needful to discuss the so-called vivisection question. I have merely to make it plain that, so far as the experiments which I am about to describe are concerned, there is **not** any reasonable ground for supposing that pain can have been suffered by the animals. And this it is easy to show; for the animals in question are so low in the scale of life, that to suppose them capable of conscious suffering would be in the highest degree unreasonable. Thus, for instance, they are considerably lower in the scale of organization than an oyster, and in none of the experiments which I have performed upon them has so much laceration of living tissue been entailed as that which is caused by opening an oyster and eating it alive, after due application of pepper and vinegar. Therefore, if any one should be foolish enough to object to my experiments on the score of vivisection, *a fortiori* they are bound to object to the culinary use of oysters. Of course, it may be answered to this that two blacks do not make a white, and that I have not by this illustration succeeded in proving my negative. To this, however, I may in turn reply that, for the purpose of morally justifying my experiments on the ground which I have adopted, it is not incumbent on me to prove any negative; it is rather for my critics to prove a positive. That is to say, before convincing me of sin, it must be shown that there is some reasonable ground for supposing that a jelly-fish or a star-fish is capable of feeling pain. I submit that there is no such ground. The mere fact that the animals are alive constitutes no such ground; for the insectivorous plants are also alive, and exhibit even more physiological "sensitiveness" and capability of rapid response to stimulation than is the case with the animals which we are about to consider. And if any one should go so far as to object to Mr. Darwin's experiments on these plants on account of its not being demonstrable that the tissues did not suffer under his operations, such a person is logically bound to go still further, and to object on similar grounds to the horrible cruelty of skinning potatoes and boiling them alive.

Thus, before any rational scruples can arise with regard to the vivisection of a living organism, some reasonable ground must be shown for supposing that the organism, besides being living, is also capable of suffering. But no such reasonable ground can be shown in the case of these low animals. We only know of such capability in any case through the analogy based upon our own experience, and, if we trust to this analogy, we must conclude that the capability in question vanishes long before we come to animals so low in the scale as the jelly-fish or star-fish. For within the limits of our own organism we have direct evidence that nervous mechanisms, much more highly elaborated than any of those which we are about to consider, are incapable of suffering. Thus, for instance, when the nervous continuity of the spinal cord is interrupted, so that a stimulus applied to the lower extremities is unable to pass upwards to the brain, the feet will be actively drawn away from a source of irritation without the man being conscious of any pain; the lower nervous centres in the spinal cord respond to the stimulation, but they do so without *feeling* the stimulus. In order to feel there must be consciousness, and, so far as our evidence goes, it appears that consciousness only arises when a nerve-centre attains to some such degree of complexity and elaboration as are to be met with in the brain. Whether or not there is a dawning consciousness in any nerve-centres considerably lower in the scale of nervous evolution, is a question which we cannot answer; but we may be quite certain that, if such is the case, the consciousness which is present must be of a commensurately dim and unsuffering kind. Consequently, even on this positive aspect of the question, we may be quite sure that by the time we come to the jelly-fish—where the object of the experiments in the first instance was to obtain evidence of the very existence of nerve-tissue—all question of pain must have vanished. Whatever opinions, therefore, we may severally entertain on the vexed question of vivisection as a whole, and with whatever feelings we may regard the "blind Fury" who, in the person of the modern physiologist, "comes with the abhorred shears and slits the thin-spun life," we should be all agreed that in the case of these animals the life is so very thin-spun that any suggestion of abhorrence is on the face of it absurd.*

* The relation of consciousness to the elaboration of nerve-centres throughout the animal kingdom is more fully considered in my work on "Mental Evolution in Animals" (Kegan Paul, Trench & Co.: 1883).

CHAPTER I.

STRUCTURE OF THE MEDUSÆ.

To give a full account of the morphology, development, and classification of the Medusæ would be both unnecessary for our present purposes and impracticable within the space which is allotted to the present work.* But, for the sake of clearness in what follows, I shall begin by briefly describing such features in the anatomy of the jelly-fish as will afterwards be found especially to concern us.

In size, the different species of Medusæ vary from that of a small pea to that of a large umbrella having streamers a hundred feet long. The general form of these animals varies in different species from that of a thimble (Fig. 1) to that of a bowl, a parasol, or a saucer (see figures in subsequent chapters). Or we may say that the form of the animals always resembles that of a mushroom, and that the resemblance

Fig. 1.—
Sarsia
(natural
size).

* Those who may desire to read an excellent epitome of our most recent knowledge on these subjects, may refer to Professor E. Ray Lankester's article in the "Encyclopædia Britannica" on "Hydrozoa," together with Professor Haeckel's Report on the Medusæ of the *Challenger* Expedition.

extends to a tolerably close imitation by different species of the various forms which are characteristic of different species of mushrooms, from the thimble-like kinds to the saucer-like kinds. Moreover, this accidental resemblance to a mushroom is increased by the presence of a central organ, occupying the position of, and more or less resembling in form, the stalk of a mushroom. This organ is called the "manubrium," on account of its looking like the "handle" of an umbrella, and the term "umbrella" is applied to the other portion of the animal. The manubrium, like the umbrella, varies much in size and shape in different species, as a glance at any figures of these animals will show. Both the manubrium and umbrella are almost entirely composed of a thick, transparent, and non-contractile jelly; but the whole surface of the manubrium and the whole *concave* surface of the umbrella are overlayed by a thin layer or sheet of contractile tissue. This tissue constitutes the earliest appearance in the animal kingdom of true muscular fibres, and its thickness, which is pretty uniform, is nowhere greater than that of very thin paper.

The manubrium is the mouth and stomach of the animal, and at the point where it is attached to or suspended from the umbrella its central cavity opens into a tube-system, which radiates through the lower or concave aspect of the umbrella. This tube-system, which serves to convey digested material and may therefore be regarded as intestinal in function, presents two different forms in the two main groups into which the Medusæ are divided. In the "naked-eyed" group, the tubes are unbranched and run in a straight course to the margin of the umbrella, where they open into a common circular tube which runs all the way round the margin (see Figs. 1 and 22). In the "covered-eyed" group, on the other hand, the tubes are strongly branched (see Fig. 8), although they likewise all eventually terminate in a single circular tube. This circular or marginal tube in both cases communicates by minute apertures with the external medium.

The margin of the umbrella, both in the naked and covered eyed Medusæ, supports a series of contractile tentacles, which vary greatly in size and number in different species (see Figs. 1 and 8). The margin also supports another series of bodies which will presently be found to be of much importance for us. These are the so-called "marginal bodies," which vary in number, size, and structure in different species. In all the covered-eyed species these marginal bodies occur in the form of little bags of crystals (therefore they are called "lithocysts"), which are protected by curiously formed "hoods" or "covers" of gelatinous tissue; and it is on this account that the group is called "covered-eyed," in contradistinction to the "naked-eyed," where these little hoods or coverings are invariably absent (compare Fig. 1 with Fig. 22), and the crystals frequently so. In nearly all cases these marginal bodies contain more or less brightly coloured pigments.

The question whether any nervous tissue is

present in the Medusæ is one which has long occupied the more or less arduous labours of many naturalists. The question attracted so much investigation on account of its being one of unusual interest in biology. Nerve-tissue had been clearly shown to occur in all animals higher in the zoological scale than the Medusæ, so that it was of much importance to ascertain whether or not the first occurrence of this tissue was to be met with in this class. But, notwithstanding the diligent application of so much skilled labour, up to the time when my own researches began there had been so little agreement in the results obtained by the numerous investigators, that Professor Huxley—himself one of the greatest authorities upon the group—thus defined the position of the matter in his " Classification of Animals " (p. 22) : " No nervous system has yet been discovered in any of these animals."

The following is a list of the more important researches on this topic up to the time which I have just named:—Ehrenberg, "Die Acalephen des rothen Meeres und der Organismvs der Medusen der Ostsee," Berlin, 1836 ; Kölliker, " Ueber die Randkörper der Quallen, Polypen und Strahlthiere," Froriep's neue Notizen, bd. xxv., 1843; Von Beneden, "Mémoire sur les Campanulaires de la côte d'Ostende," " Mémoires de l'Académie de Bruxelles," vol. xvii., 1843; Desor, " Sur la Génération Medusipare des Polypes hydraires," "Annales d. Scienc. Natur. Zool.," ser. iii. t. xii. p. 204 ; Krohn, " Ueber Podocoryna carnea,"

" Archiv. f. Naturgeschichte," 1851, b. i.; McCrady,
" Descriptions of Oceania, etc.," " Proceedings of the
Elliot Society of Natural History," vol. i., 1859;
L. Agassiz, "Contributions to the Acaliphæ of
North America," "Memoirs of the American
Academy of Arts and Sciences," vol. iii., 1860,
vol. iv., 1862; Leuckart, "Archiv. f. Naturge-
schichte," Jahrg. 38, b. ii., 1872; Hensen, " Studien
über das Gehörorgan der Decapoden," "Zeitchr. f.
wiss. Zool.," bd. xiii., 1863; Semper, " Reisebericht,"
" Zeitschr. f. wiss. Zool.," bd. xiii. vol. xiv.; Claus,
" Bemerkungen über Ctenophoren und Medusen,"
"Zeitschr. f. wiss. Zool.," bd. xiv., 1864; Allman,
"Note on the Structure of Certain Hydroid
Medusæ," "Brit. Assoc. Rep.," 1867; Fritz Müller,
" Polypen und Quallen von S. Catherina," " Archiv.
f. Naturgesch.," Jahrg. 25, bd. i., 1859; also " Ueber
die Rand-bläschen der Hydroidquallen," "Archiv.
f. Anatomie und Physiologie," 1852; Haeckel,
"Beiträge zur Naturgesch. der Hydromedusen,"
1865; Eimer, "Zoologische Untersuchungen," Würz-
burg, " Verhandlungen der Phys.-med. Gesellschaft,"
N.F. vi. bd., 1874.

The most important of these memoirs for us to
consider are the two last. I shall subsequently
consider the work of Dr. Eimer, which up to this
date was of a purely physiological character. Professor Haeckel, who made his microscopical observations chiefly upon the Geryonidæ, described the
nervous elements as forming a continuous circle all
round the margin of the umbrella, following the
course of the radial or nutrient tubes throughout

their entire length, and proceeding also to the tentacles and marginal bodies. At the base of each tentacle there is a ganglionic swelling, and it is from these ganglionic swellings that the nerves just mentioned take their origin. The most conspicuous of these nerves are those that proceed to the radial canals and marginal bodies, while the least conspicuous are those that proceed to the tentacles. Cells, as a rule, can only be observed in the ganglionic swellings, where they appear as fusiform and distinctly nucleated bodies of great transparency and high refractive power. On the other hand, the nerves that emanate from the ganglia are composed of a delicate and transparent tissue, in which no cellular elements can be distinguished, but which is longitudinally striated in a manner very suggestive of fibrillation. Treatment with acetic acid, however, brings out distinct nuclei in the case of the nerves that are situated in the marginal vesicles, while in those that accompany the radial canals ganglion-cells are sometimes met with.

A brief sketch of the contents of these and other memoirs on the histology of the Medusæ is given by Drs. Hertwig in their more recently published work on the nervous system and sense-organs of the Medusæ, and these authors point to the important fact that before the appearance of Haeckel's memoir, Leuckart was the only observer who spoke for the fibrillar character of the so-called marginal ring-nerve; so that in Haeckel's researches on Geryonia, whereby both true ganglion-cells and true nerve-fibres were first demonstrated as occurring in the

Medusæ, we have a most important step in the histology of these animals. Haeckel's results in these respects have since been confirmed by Claus, "Grundzüge der Zoologie," 1872; Allman, "A Monograph of the Gymnoblastic or Tubularian Hydroids," 1871; Harting, "Notices Zoologiques," Niedlandisches "Archiv. f. Zool.," bd. ii., Heft 3, 1873; F. E. Schulze, "Ueber den Bau von Syncorzne Sarsii"; O. and R. Hertwig, "Das Nervensystem und die Sinnesorgane der Medusen."

The last-named monograph is much the most important that has appeared upon the histology of the Medusæ. I shall, therefore, give a condensed epitome of the leading results which it has established.

There is so great a difference between the nervous system of the naked and of the covered eyed Medusæ, that a simultaneous description of the nervous system in both groups is not by these authors considered practicable. Beginning, therefore, with the naked-eyed division, they describe the nervous system as here consisting of two parts, a central and a peripheral. The central part is localized in the margin of the swimming-bell, and there forms a "nerve-ring," which is divided by the insertion of the "veil"* into an upper and a lower nerve-ring. In many species the upper nerve-ring is spread out in the form of a flattish layer, which

---

* This is the name given to a small annular sheet of tissue which forms a kind of floor to the orifice of the swimming-bell, through the central opening of which floor the manubrium passes. The structure is shown in Fig. 1.

is somewhat thickened where it is in contact with the veil. In these species the nerve-ring is only indistinctly marked off from the surrounding tissues. But in other species the crowding together of the nerve-fibres at the insertion of the veil gives rise to a considerable concentration of nervous structures; while in others, again, this concentration proceeds to the extent of causing a well-defined swelling of nervous tissue against the epithelium of the veil and umbrella. In the Geryonidæ this swelling is still further strengthened by a peculiar modification of the other tissues in the neighbourhood, which had been previously described by Professor Haeckel. In all species the upper nerve-ring lies entirely in the ectoderm. Its principal mass is composed of nerve-fibres of wonderful tenuity, among which are to be found sparsely scattered ganglion-cells. The latter are for the most part bi-polar, more seldom multi-polar. The fibres which emanate from them are very delicate, and, becoming mixed with others, do not admit of being further traced. Where the nervous tissue meets the enveloping epithelium it is connected with the latter from within, but differs widely from it; for the nerve-cells contain a longitudinally striated cylindrical or thread-like nucleus which carries on its peripheral end a delicate hair, while its central end is prolonged into a fine nerve-fibre. There are, besides these, two other kinds of cells which form a transition between the ganglion and the epithelium cells. The first kind are of a long and cylindrical form, the free ends of which reach as far as the upper surface of the epithelium

The second kind lie for the most part under the upper surface. They are of a large size, and present, coursing towards the upper surface, a long continuation, which at its free extremity supports a hair. In some cases this continuation is smaller, and stops short before reaching the outer surface. Drs. Hertwig observe that in these peculiar cells we have tissue elements which become more and more like the ordinary ganglion-cells of the nerve-ring the more that their long continuation towards the surface epithelium is shortened or lost, and these authors are thus led to conclude that the upper nerve-ring was originally constituted only by such prolongations of the epithelium-cells, and that afterwards these prolongations gradually disappeared, leaving only their remnants to develop into the ordinary ganglion-cells already described.

Beneath the upper nerve-ring lies the lower nerve-ring. It is inserted between the muscle-tissue of the veil and umbrella, in the midst of a broad strand wherein muscle-fibres are entirely absent. It here constitutes a thin though broad layer which, like the upper nerve-ring, belongs to the ectoderm. It also consists of the same elements as the upper nerve-ring, viz. of nerve-fibres and ganglion-cells. Yet there is so distinct a difference of character between the elements composing the two nerve-rings, that even in an isolated portion it is easy to tell from which ring the portion has been taken. That is to say, in the lower nerve-ring there are numerous nerve-fibres of considerable thickness, which contrast in a striking manner with the almost immeasurably slender fibres of the upper nerve-ring. A second point of difference consists in the surprising wealth of ganglion-cells in the one ring as compared with the other. Thus, on the whole, there is no doubt that the lower nerve-ring presents a higher grade of structure than does the upper, as shown not only by the greater multiplicity of nerve-cells and fibres, but also by the relation in which these elements stand to the epithelium. For in the case of the lower nerve-ring, the presumably primitive connections of the nervous elements with the epithelium is well-nigh dissolved—this nerve-ring having thus separated itself from its parent structure, and formed for itself an independent layer beneath the epithelium. The two nerve-rings are separated from one another by a very thin membrane, which, in some species at all events, is bored through by strands of nerve-fibres which serve to connect the two nerve-rings with one another.

The peripheral nervous system is also situated in the ectoderm, and springs from the central nervous system, not by any observable nerve-trunks, but directly as a nervous plexus composed both of cells and fibres. Such a nervous plexus admits of being detected in the sub-umbrella of all Medusæ, and in some species may be traced also into the tentacles. It invariably lies between the layer of muscle-fibre and that of the epithelium. The processes of neighbouring ganglion-cells in the plexus either coalesce or dwindle in their course to small fibres: at the margin of the umbrella these unite themselves with the elements of the nerve-rings. There are also described several peculiar tissue elements, such as, in the umbrella, nerve-fibres which probably stand in connection with epithelium-cells; nerve-cells which pass into muscle-fibres, similar to those which Kleinenberg has called neuro-muscular cells; and, in the tentacles, neuro-muscular cells joined with cells of special sensation (*Sinneszellen*).

No nervous elements could be detected in the convex surface of the umbrella, and it is doubtful whether they occur in the veil.

In some species the nerve-fibres become aggregated in the region of the generative organs, and in that of the radial canals, thus giving rise in these localities to what may be called nerve-trunks. But in other species no such aggregations are apparent, the nervous plexus spreading out in the form of an even trellis-work.

In the covered-eyed Medusæ the central nervous system consists of a series of separate centres which are not connected by any commissures. These nerve-centres are situated in the margin of the umbrella, and are generally eight in number, more rarely twelve, and in some species sixteen. They are thickenings of the ectoderm, which either enclose the bases of the sense-organs, or only cover the ventral side of the same. Histologically they consist of cells of special sensation, together with a thick layer of slender nerve-fibres. Ganglion-cells, however, are absent, so that the nerve-fibres are merely processes of epithelium-cells.

Drs. Hertwig made no observations on the peripheral nervous system of the covered-eyed Medusæ; but they do not doubt that such a system would admit of being demonstrated, and in this connection they cite the observations of Claus, who describes numerous ganglion-cells as occurring in the sub-umbrella of Chrysaora. Here I may appropriately state that before Drs. Hertwig had published their results, Professor Schäfer, F.R.S., conducted in my laboratory a careful research upon the histology of the Medusæ, and succeeded in showing an intricate plexus of cells and fibres overspreading the sub-umbrella tissue of another covered-eyed Medusa (Aurelia aurita).* He also found that the marginal bodies present a peculiar modification of epithelium tissue, which is on its way, so to speak, towards becoming fully differentiated into ganglionic cells.

Lastly, returning to the researches of Drs. Hertwig, these authors compare the nervous system of the naked-eyed with that of the covered-eyed Medusæ, with the view of indicating the points which show the latter to be less developed than the former. These points are, that in the nerve-centres of the covered-eyed Medusæ there are no true ganglion-cells, or only very few; that the mass of the central nervous system is very small; and that the centralization of the nervous system is less complete in the one group than in the other. In their memoir these authors further supply much

* See "Observations on the Nervous System of *Aurelia aurita*," *Phil. Trans.*, pt. ii., 1878.

interesting information touching the structure of the sense-organs in various species of Medusæ; but it seems scarcely necessary to extend the present *résumé* of their work by entering into this division of their subject.

In a later publication, entitled "Der Organismus der Medusen und seine Stellung zur Keimblätter-theorie," Drs. Hertwig treat of sundry features in the morphology of the Medusæ which are of great theoretical importance; but here again it would unduly extend the limits of the present treatise if I were to include all the ground which has been so ably cultivated by these industrious workers.

It will presently be seen in how striking a manner all the microscopical observations to which I have now briefly alluded are confirmed by the physiological observations—or, more correctly, I might say that the microscopical observations, in so far as they were concerned with demonstrating the existence of nerve-tissue in the Medusæ, were forestalled by these physiological experiments; for, with the exception of Professor Haeckel's work on Geryonidæ, they were all of later publication. But in matters of scientific inquiry mere priority is not of so much importance as it is too often supposed to be. Thus, in the present instance, no one of the workers was in any way assisted by the publications of another. In each case the work was independent and almost simultaneous.

The remark just made applies also to the only research which still remains to be mentioned. This is the investigation undertaken and published by

Professor Eimer.* He began, like myself, by what in the next chapter I call the "fundamental observation" on the effects of excising the nerve-centres, and from this basis he worked both at the physiology and the morphology of the neuro-muscular tissues. In point of time, I was the first to make the fundamental observation, and he was the first to publish it. The sundry features in which our subsequent investigations agreed, and those in which they differed, I shall mention throughout the course of the following pages.

I shall now conclude this chapter by giving a brief account of those general principles of the physiology of nerve and muscle with which it is necessary to be fully acquainted, in order to understand the course of the following experiments.

Nerve-tissue, then, universally consists of two elementary structures, viz. very minute nerve-cells and very minute nerve-fibres. The fibres proceed to and from the cells, so in some cases serving to unite the cells with one another, and in other cases with distant parts of the animal body. Nerve-cells are usually found collected together in aggregates, which are called nerve-centres or ganglia, to and from which large bundles of nerve-fibres come and go.

To explain the *function* of nerve-tissue, it is necessary to begin by explaining what physiologists mean by the term "excitability." Suppose that a

* "Die Medusen physiologisch und morphologisch auf ihr Nervensystem untersucht" (Tübingen, 1878).

muscle has been cut from the body of a freshly killed animal; so long as it is not interfered with in any way, so long will it remain quite passive. But every time a stimulus is supplied to it, either by means of a pinch, a burn, an electrical shock, or a chemical irritant, the muscle will give a single contraction in response to every stimulation. And it is this readiness of organic tissues to respond to a suitable stimulus that physiologists designate by the term "excitability."

Nerves, no less than muscles, present the property of being excitable. If, together with the excised muscle, there had been removed from the animal's body an attached nerve, every time any part of this nerve is stimulated the attached muscle will contract as before. But it must be carefully observed that there is this great difference between these two cases of response on the part of the muscle—that while in the former case the muscle responded to a stimulus *applied directly to its own substance,* in the latter case the muscle responded to a stimulus applied *at a distance from its own substance,* which stimulus was then *conducted* to the muscle by the nerve. And in this we perceive the characteristic function of nerve-*fibres,* viz. that of conducting stimuli to a distance. The function of nerve-*cells* is different, viz. that of accumulating nervous energy, and, at fitting times, of discharging this energy into the attached nerve-fibres. The nervous energy, when thus discharged, acts as a stimulus to the nerve-fibre; so that if a muscle is attached to the end of a fibre, it contracts on receiving this stimulus. I may add that when nerve-cells are collected into ganglia, they often appear to discharge their energy spontaneously; so that in all but the very lowest animals, whenever we see apparently *spontaneous* action, we infer that ganglia are probably present. Lastly, another important distinction must be borne in mind—the distinction, namely, which is to be drawn between muscle and nerve. A stimulus applied to a nerveless muscle can only course through the muscle by giving rise to a visible wave of contraction, which spreads in all directions from the seat of disturbance as from a centre. A nerve, on the other hand, conducts the stimulus without sensibly moving or undergoing any change of shape. Now, in order not to forget this distinction, I shall always speak of muscle-fibres as conveying a *visible* wave of *contraction*, and of nerve-fibres as conveying an *invisible*, or *molecular*, wave of *stimulation*. Nerve-fibres, then, are functionally distinguished from muscle-fibres—and also from protoplasm—by displaying the property of conducting invisible, or molecular, waves of stimulation from one part of an organism to another, so establishing physiological continuity between such parts, *without the necessary passage of waves of contraction.*

CHAPTER II.

FUNDAMENTAL EXPERIMENTS.

THE naked-eyed Medusæ are very much smaller in size than the covered-eyed, and as we shall find that the distribution of their nervous elements is somewhat different, it will be convenient to use different names for the large umbrella-shaped part of a covered-eyed Medusa, and the much smaller though corresponding part of a naked-eyed Medusa. The former, therefore, I shall call the umbrella, and the latter the swimming-bell, or nectocalyx. In each case alike this portion of the animal performs the office of locomotion, and it does so in the same way. I have already said that this mushroom-like organ, which constitutes the main bulk of the animal, is itself mainly constituted of thick transparent and non-contractile jelly, but that the whole of its concave surface is lined with a thin sheet of muscular tissue. Such being the structure of the organ, the mechanism whereby it effects locomotion is very simple, consisting merely of an alternate contraction and relaxation of the entire muscular sheet which lines the cavity of the bell. At each contraction of this muscular sheet the gelatinous walls of the bell are drawn together; the capacity of the bell being thus diminished, water is ejected from the open mouth of the bell backwards, and the consequent reaction propels the animal forwards. In these swimming movements, systole and diastole follow one another with as perfect a rhythm as they do in the beating of a heart.

Effects of excising the entire Margins of Nectocalyces.

Confining our attention under this heading to the naked-eyed Medusæ, I find that the following proposition applies to every species of the group which I have as yet had the opportunity of examining: *Excision of the extreme margin of a nectocalyx causes immediate, total, and permanent paralysis of the entire organ.* Nothing can possibly be more definite than in this highly remarkable effect. I have made hundreds of observations upon various species of the naked-eyed Medusæ, of all ages and conditions of freshness, vigour, etc.; and I have constantly found that if the experiment be made with ordinary care, so as to avoid certain sources of error presently to be named, the result is as striking and decided as it is possible to desire.* Indeed, I do not know of any case in the animal kingdom where the removal of a centre of spontaneity causes so sudden and so complete a paralysis

* I have only met with one individual exception. This occurred in a specimen of *Staurophora laciniata*, where, after removal of the entire margin, three centres of spontaneity were found to remain in the sheet of contractile tissue lining the nectocalyx.

of the muscular system, there being no subsequent movements or twitchings of a reflex kind to disturb the absolute quiescence of the mutilated organism. The experiment is particularly beautiful if performed on Sarsia; for the members of this genus being remarkably active, the death-like stillness which results from the loss of so minute a portion of their substance is rendered by contrast the more surprising.

From this experiment, therefore, I conclude that in the margin of all the species of naked-eyed Medusæ which I have as yet had the opportunity of examining, there is situated an intensely localized system of centres of spontaneity, having at least for one of its functions the origination of impulses, to which the contractions of the nectocalyx, under ordinary circumstances, are exclusively due. And this obvious deduction is confirmed (if it can be conceived to require confirmation) by the behaviour of the severed margin. This continues its rhythmical contractions with a vigour and a pertinacity not in the least impaired by its severance from the main organism, so that the contrast between the perfectly motionless swimming-bell and the active contractions of the thread-like portion which has just been removed from its margin is as striking a contrast as it is possible to conceive. Hence it is not surprising that if the margin be left *in situ*, while other portions of the swimming-bell are mutilated to any extent, the spontaneity of the animal is not at all interfered with. For instance, if the equator of any individual belonging to the genus Sarsia (Fig. 1) be cut completely through, so that the swimming-bell instead of being closed at the top is converted into an open tube, this open tube continues its rhythmical contractions for an indefinitely long time, notwithstanding that the organism so mutilated is, of course, unable to progress. Thus it is a matter of no consequence how small or how large a portion of contractile tissue is left adhering to the severed margin of the swimming-bell; for whether this portion be large or small, the locomotor centres contained in the margin are alike sufficient to supply the stimulus to contraction. Indeed, if only the tiniest piece of contractile tissue be left adhering to a single marginal body cut out of the bell of Sarsia, this tiny piece of tissue, in this isolated state, will continue its contractions for hours, or even for days.

Effects of excising the entire Margins of Umbrellas.

Turning now to the covered-eyed division of the Medusæ, I find, in all the species I have come across, that excision of the margins of umbrellas produces an effect analogous to that which is produced by excision of the margins of swimming-bells. There is an important difference, however, between the two cases, in that the paralyzing effect of the operation on umbrellas is neither so certain nor so complete as it is on swimming-bells. That is to say, although in the majority of experiments such mutilation of umbrellas is followed by immediate paralysis, this is not invariably the case; so that one cannot here, as with the naked-eyed Medusæ, predict with any great confidence what will be the immediate result of any particular experiment. Further, although such mutilation of an umbrella is usually followed by a paralysis as sudden and marked as that which follows such mutilation of a swimming-bell, the paralysis of the former differs from the paralysis of the latter, in that it is very seldom *permanent*. After periods varying from a few seconds to half an hour or more, occasional weak and unrhythmical contractions begin to manifest themselves, or the contractions may even be resumed with but little apparent change in their character and frequency. The condition of the animal before the operation, as to general vigour, etc., appears to be one factor in determining the effect of the operation; but this is very far from being the only factor.

Upon the whole, then, although in the species of covered-eyed Medusæ which I have as yet had the opportunity of examining, the effects which result from excising the margins of umbrellas are such as to warrant me in saying that the main supply of locomotor centres appears to be usually situated in that part of these organs, these effects are nevertheless such as to compel me at the same time to conclude that the locomotor centres of the covered-eyed Medusæ are more diffused or segregated than are those of the naked-eyed Medusæ. Lastly, it should be stated that all the species of covered-eyed Medusæ resemble all the species of naked-eyed Medusæ, in that their members will endure any amount of section it is possible to make upon any of their parts other than their margins without their spontaneity being in the smallest degree affected.

Effects of excising Certain Portions of the Margins of Nectocalyces.

The next question which naturally presents itself is as to whether the locomotor centres are equally distributed all round the margin of a swimming organ, or situated only, or chiefly, in the so-called marginal bodies. To take the case of the naked-eyed Medusæ first, it is evident that in most of the genera, in consequence of the intertentacular spaces being so small, it is impossible to cut out the marginal bodies (which are situated at the bases of the tentacles) without at the same time cutting out the intervening portions of the margin. The genus Sarsia, however, is admirably adapted (as a glance at Fig. 1 will show) for trying the effects of removing the marginal bodies without injuring the rest of the margin, and *vice versâ*. The results of such experiments upon members of this genus are as follow.

Whatever be the condition of the individual operated upon as to freshness, vigour, etc., it endures excision of three of its marginal bodies without suffering any apparent detriment; but in most cases, as soon as the last marginal body is cut out, the animal falls to the bottom of the water quite motionless. If the subject of the experiment happens to be a weakly specimen, it will, perhaps, never move again : it has been killed by something very much resembling nervous shock. On the other hand, if the specimen operated upon be one which is in a fresh and vigorous state, its period of quiescence will probably be but short; the nervous shock, if we may so term it, although evidently considerable at the time, soon passes away, and the animal resumes its motions as before. In the great majority of cases, however, the activity of these motions is conspicuously diminished.

The effect of excising all the marginal tissue from between the marginal bodies and leaving the latter untouched, is not so definite as is the effect of the converse experiment just described. Moreover, allowance must here be made for the fact that in this experiment the principal portion of the " veil " *
is of necessity removed, so that it becomes impossible to decide how much of the enfeebling effect of the section is due to the removal of locomotor centres from the swimming-bell, and how much to a change in the merely mechanical conditions of the organ. From the fact, however, that excision of the entire margin of Sarsia produces total paralysis, while excision of the marginal bodies alone produces merely partial paralysis, there can be no doubt that both causes are combined. Indeed, it has been a matter of the greatest surprise to me how very minute a portion of the intertentacular marginal tissue is sufficient, in case of this genus, to animate the entire swimming-bell. Choosing vigorous

* See Fig. 1.

FUNDAMENTAL EXPERIMENTS.

specimens of Sarsia, I have tried, by cutting out all the margin besides, to ascertain how minute a portion of intertentacular tissue is sufficient to perform this function, and I find that this portion may be so small as to be quite invisible without the aid of a lens.

From numerous observations, then, upon Sarsia, I conclude that in this genus (and so, from analogy, probably in all the other genera of the true Medusæ) locomotor centres are situated in every part of the extreme margin of a nectocalyx, but that there is a greater supply of such centres in the marginal bodies than elsewhere.

Effects of excising Certain Portions of the Margin of Umbrellas.

Coming now to the covered-eyed Medusæ, I find that the concentration of the locomotor centres of the margin into the marginal bodies, or lithocysts, is still more decided than it is in the case of Sarsia. Taking Aurelia aurita as a type of the group, I cannot say that, either by excising the lithocysts alone or by leaving the lithocysts *in situ* and excising all the rest of the marginal tissue, I have ever detected the slightest indications of locomotor centres being present in any part of the margin of the umbrella other than the eight lithocysts; so that all the remarks previously made upon this species, while we were dealing with the effects of excising the entire margin of umbrellas, are equally applicable to the experiment we are now considering, viz. that of excising the lithocysts alone. In other words, but for the sake of symmetry, I might as well have stated at the first that in the case of the covered-eyed Medusæ all the remarkable paralyzing effects which are obtained by excising the entire margin of an umbrella are obtained in exactly the same degree by excising the eight lithocysts alone; the intermediate marginal tissue, in the case of these Medusæ, is totally destitute of locomotor centres.

Effects upon the Manubrium of excising the Margin of a Nectocalyx or Umbrella.

Lastly, it must now be stated, and always borne in mind, that neither in the case of naked nor covered eyed Medusæ does excision of the margin of a swimming organ produce the smallest effect upon the manubrium. For hours and days after the former, in consequence of this operation, has ceased to move, the latter continues to perform whatever movements are characteristic of it in the unmutilated organism—indeed, these movements are not at all interfered with even by a complete severance of the manubrium from the rest of the animal. In many of the experiments subsequently to be detailed, therefore, I began by removing the manubrium, in order to afford better facilities for manipulation.

Summary of Chapter.

With a single exception to hundreds of observations upon six widely divergent genera of naked-eyed Medusae, I find it to be uniformly true that removal of the extreme periphery of the animal causes instantaneous, complete, and permanent paralysis of the locomotor system. In the genus Sarsia, my observations point very decidedly to the conclusion that the principal locomotor centres are the marginal bodies, but that, nevertheless, every microscopical portion of the intertentacular spaces of the margin is likewise endowed with the property of originating locomotor impulses.

In the covered-eyed division of the Medusæ, I find that the *principal* seat of spontaneity is the margin, but that the latter is not, as in the naked-eyed Medusæ, the *exclusive* seat of spontaneity. Although in the vast majority of cases I have found that excision of the margin impairs or destroys the spontaneity of the animal for a time, I have also found that the paralysis so produced is very seldom of a permanent nature. After a variable period occasional contractions are usually given, or, in some cases, the contractions may be resumed with but little apparent detriment. Considerable differences, however, in these respects are manifested by different species, and also by different individuals of the same species. Hence, in comparing the covered-eyed group as a whole with the naked-eyed group as a whole, so far as my observations extend, I should say that the former resembles the latter in that its representatives usually have their main supply of locomotor centres situated in their margins, but that it differs from the latter in that its representatives usually have a greater or less supply of their locomotor centres scattered through the general contractile tissue of their swimming organs. But although the locomotor centres of a covered-eyed Medusa are thus, generally speaking, more diffused than are those of a naked-eyed Medusa, *if we consider the organism as a whole*, the locomotor centres in the *margin* of a covered-eyed Medusa are *less* diffused than are those in the *margin* of a naked-eyed Medusa. In no case does the excision of the margin of a swimming organ produce any effect upon the movements of the manubrium.

CHAPTER III.

EXPERIMENTS IN STIMULATION.

Mechanical, Chemical, and Thermal Stimulation.

So far as my observations extend, I find that all Medusæ, after removal of their locomotor centres, invariably respond to every kind of stimulation. To take the case of Sarsia as a type, nothing can possibly be more definite than is the single sharp contraction of the mutilated nectocalyx in response to every nip with the forceps. The contraction is precisely similar to the ordinary ones that are performed by the unmutilated animal; so that by repeating the stimulus a number of times, the nectocalyx, with its centres of spontaneity removed, may be made to progress by a succession of contractions round and round the vessel in which it is contained, just as a frog, with its cerebral hemispheres removed, may be made to hop along the table in response to a succession of stimulations.*

* In the case of the covered-eyed Medusæ, however, the paralyzed umbrella sometimes responds to a single stimulation with two, and more rarely with three contractions, which are separated from one another by an interval of the same duration as the normal diastole of the unmutilated animal.

Different species of Medusæ exhibit different degrees of irritability in responding to stimuli; but in all the cases I have met with the degree of irritability is remarkably high. Thus, I have seen responsive contractions of the whole umbrella follow upon the exceedingly slight stimulus caused by a single drop of sea-water let fall upon the irritable surface from the height of one inch. As regards chemical stimulation, dilute spirit or other irritant, when dropped on the paralyzed swimming organ of Aurelia aurita, often gives rise to a whole series of rhythmical pulsations, the systoles and diastoles following one another at about the same rate as is observable in the normal swimming motions of the unmutilated animal.

It is somewhat difficult, in the case of paralyzed swimming organs, to prove the occurrence of a contraction in respon e to thermal stimulation, from the fact that while these tissues are not nearly so sensitive to this mode of excitation as might be anticipated, they are, as just observed, extraordinarily sensitive to mechanical excitation. It therefore becomes difficult to administer the appropriate thermal stimulus without at the same time causing a sufficient mechanical disturbance to render it doubtful to which of the stimuli the response is due. This may be done, however, by allowing a few drops of heated sea-water to run over the excitable surface while it is exposed to the air. In this and in other ways I have satisfied myself that the paralyzed tissues of swimming organs respond to sudden elevations of temperature.

Luminous Stimulation.

It is interesting to note that, in the case of some of the naked-eyed Medusæ, the action of light as a stimulus is most marked and unfailing. In the case of Sarsia, for instance, a flash of light let fall upon a living specimen almost invariably causes it to respond with one or more contractions. If the animal is vigorous and swimming freely in water, the effect of a momentary flash thrown upon it during one of the natural pauses is immediately to originate a bout of swimming. But if the animal is non-vigorous, or if it be removed from the water and spread flat upon an object-glass, it usually gives only one contraction in response to every flash. There can thus be no doubt that a sudden transition from darkness to light acts upon Sarsia as a stimulus, and this even though the transition be but of momentary duration. The question therefore arises as to whether the stimulus consists in the presence of light, or in the occurrence of the sudden transition from darkness to light and from light to darkness. To answer this question, I tried the converse experiment of placing a vigorous specimen in sunlight, waiting till the middle of one of the quiescent stages in the swimming motions had come on, and then suddenly darkening. In no case, however, under these circumstances, did I obtain any response; so that I cannot doubt it is the light *per se*, and not the sudden nature of the transition from darkness to light, which in the former experiment acted as the stimulus. Indeed, the effect of the converse experiment just described is rather that of inhibiting contractions; for, if the sunlight be suddenly shut off during the occurrence of a swimming bout, it frequently happens that the quiescent stage immediately sets in. Again, in a general way, it is observable that Sarsiæ are more active in the light than they are in the dark, the comparative duration of the quiescent stages being less in the former than in the latter case. Light thus appears to act towards these animals as a constant stimulus. Lastly, it may be stated that when the marginal bodies of Sarsia are removed, the swimming-bell, although still able to contract spontaneously, no longer responds to luminous stimulation of any kind or degree. But if only one body be left *in situ*, or if the severed margin alone be experimented upon, the same unfailing response may be obtained to luminous stimulation as that which is obtained from the entire animal.

The fact last mentioned indicates that the marginal bodies are organs of special sense, adapted to respond to luminous stimulation; or, in more simple words, that they perform the functions of sight. Now it has long been thought more or less probable that these marginal bodies are rudimentary or incipient "eyes," but hitherto the supposition has not been tested by experiment, and was therefore of no more value than a guess.* The guess in this

---

* As Professor Haeckel observes in his monograph already alluded to, " Die Deutung der Sinnesorgane niederer Thiere gehört ohne Zweifel zu den schwierigsten Objecten der vergleichenden Physiologie und ist der grössten Unsicherheit unterworfen. Wir sind gewohnt, die von den Wirbelthieren gewonnenen

instance, however, happens to have been correct, as he results of the following experiments will show.

Having put two or three hundred Sarsiæ into a large bell-jar, I completely shut out the daylight from the room in which the jar was placed. By means of a dark lantern and a concentrating lens, I then cast a beam of light through the water in which the Sarsiæ were swimming. The effect upon the latter was most decided. From all parts of the bell-jar they crowded into the path of the beam, and were most numerous at that side of the jar which was nearest to the light. Indeed, close against the glass they formed an almost solid mass, which followed the light wherever it was moved. The individuals composing this mass dashed themselves against the glass nearest the light with a vigour and determination closely resembling the behaviour of moths under similar circumstances. There can thus be no doubt about Sarsia possessing a visual sense.

Anschauungen ohne Weiteres auch auf die wirbellosen Thiere der verschiedenen Kirese zu übertragen und bei diesen analoge Sinnesempfindungen anzunehmen als wir sebst besitzen... Noch weniger freilich als die von den meisten Autoren angenommene Deutung der Randbläschen unserer Medusen als Gehörorgane kann die von Agassiz und Fritz Müller vertretene Ansicht befriedigen, dass dieselben Augen seien.... Alle diese Verhältnisse sind mit der Deutung der Concretion als 'Linse' und des sie umschliessenden Sinnesganglion als 'Sehnerv' durchaus unvereinbar."

It may not be unnecessary to say that, although the simple experiment above described effectually proves that the marginal bodies have a visual function to subserve, we are not for this reason justified in concluding that these are so far specialized as organs of sight as to be precluded from ministering to any other sense.

The method of ascertaining whether this sense is lodged in the marginal bodies was, of course, extremely simple. Choosing a dozen of the most vigorous specimens, I removed all the marginal bodies from nine, and placed these, together with the three unmutilated ones, in another bell-jar. After a few minutes the mutilated animals recovered from their nervous shock, and began to swim about with tolerable vigour. I now darkened the room, and threw the concentrated beam of light into the water as before. The difference in the behaviour of the mutilated and of the unmutilated specimens was very marked. The three individuals which still had their marginal bodies sought the light as before, while the nine without their marginal bodies swam hither and thither, without paying it any regard.

A further question, however, still remained to be determined. The pigment spot of the marginal body in Medusæ is, as L. Agassiz observed, placed *in front of* the presumably nervous tissue, and for this reason he naturally enough suggested that if the marginal body has a visual function to perform, the probability is that the rays by which the organ is affected are the heat-rays lying beyond the range of the visible spectrum. Accordingly I brought a heated iron, just ceasing to be red, close against the large bell-jar which contained the numerous specimens of Sarsia; but not one of the latter approached the heated metal.

From these observations, therefore, I conclude that in Sarsia the faculty of appreciating luminous rays is present, and that this faculty is lodged exclusively in the marginal bodies; while from observations conducted on the covered-eyed Medusæ, I have come to the same conclusion respecting them. But although I have tested many species of naked-eyed Medusæ besides Sarsia, I have obtained indications of response to luminous stimulation only in the case of one other. This is a species which I have called Tiaropsis polydiademata, and the response which it gives to luminous stimulation is even more marked and decided than that which is given by Sarsia; for a sudden exposure to sunlight causes this animal to go into a kind of tonic spasm, the whole of the nectocalyx being drawn together in a manner resembling cramp. Now, in one remarkable particular this response to luminous stimulation on the part of Tiaropsis polydiademata differs from that given by Sarsia tubulosa; and the difference consists in the fact that, while with Sarsia the period of latency (*i.e.* the time between the fall of the stimulus and the occurrence of the response) is, so far as the eye can judge, as instantaneous in the case of response to luminous stimulation as it is in the case of response to any other kind of stimulation, such is far from being true with Tiaropsis polydiademata. The period of latency in the last-named species is, so far as the eye can judge, quite as instantaneous as it is in the case of Sarsia, when the stimulus employed is other than luminous; but in response to light, the characteristic spasm does not take place till slightly more than a second has elapsed after the first occurrence of the stimulus. As this extraordinary difference in the latent period exhibited by the same animal towards different kinds of stimuli appeared to me a matter of considerable interest, I was led to reflect upon the probable cause of the difference. It occurred to me that the only respect in which luminous stimulation of the Medusæ differed from all the other modes of stimulation I had employed consisted in this—that, as proved by my previous experiments on Sarsia, which I repeated on Tiaropsis, luminous stimulation directly affected the ganglionic tissues. Now, as in Tiaropsis polydiademata luminous stimulation differed from all the other modes of stimulation in giving rise to an immensely longer period of latency, I seemed here to have an index of the difference between the rapidity of the response to stimuli by the contractile and by the ganglionic tissues respectively. The next question, therefore, which presented itself was as to whether the enormous length of time occupied by the process of stimulation in the ganglia was due to any necessity on the part of the latter to accumulate the stimulating influence prior to originating a discharge, or to an immensely lengthened period of latent stimulation manifested by the ganglia under the influence of light.* This is an

* The period of latent stimulation merely means the time after the occurrence of an excitation during which a series of physiological processes are taking place, which terminate in a contraction; so that, whether the excitation is of a strong or of a weak intensity, the period of latent stimulation is not much affected. The above question, therefore, was simply this—Does the prolonged delay on the part of these ganglia, in responding to light,

interesting question, because if such a lengthened period of latent stimulation occurs in this case, it would stand in curious antithesis to the very short period of latent stimulation manifested by the contractile tissues of the same animal under other modes of irritation. To test these alternative hypotheses, I employed the very simple method of first allowing a continuous flood of light to fall suddenly on the Medusa, and then noting the time at which the responsive spasm first began. This time, as already stated, was slightly more than one second. I next allowed the animal to remain for a few minutes in the dark to recover shock, and, lastly, proceeded to throw in single flashes of light of measured duration. I found that unless the flash of light was of slightly more than one second in its duration, no response was given; that is to say, the minimal duration of a flash required to produce a responsive spasm was just the same as the time during which a continuous flood of light required to operate in order to produce a similar spasm. From this, therefore, I conclude that the enormously long period of latent excitation in response to luminous stimuli was not, properly speaking, a period of latent excitation at all; but that it represented the time during which a certain summation of stimulating influence was taking place in the ganglia, which required somewhat more

represent the time during which the series of physiological processes are taking place in response to an adequate stimulus, or does it represent the time during which light requires to act before it becomes an adequate stimulus?

than a second to accumulate, and which then caused the ganglia to originate an abnormally powerful discharge. So that in the action of light upon the ganglionic matter of this Medusa we have some analogy to its action on certain chemical compounds in this respect, that, just as in the case of those compounds which light is able to split up, a more or less lengthened exposure to its influence is necessary in order to admit of the summating influence of its vibrations on the molecules, so in the case of this ganglionic material, the decomposition which is effected in it by light, and which terminates in an explosion of nervous energy, can only be effected by a prolonged exposure of the unstable material to the summating influence of the luminous vibrations. Probably, therefore, we have here the most rudimentary type of a visual organ that is possible; for it is evident that if the ganglionic matter were a very little more stable than it is, it would altogether fail to be thrown down by the luminous vibrations, or would occupy so long a time in the process that the visual sense would be of no use to its possessor. How great is the contrast between the excitability of such a sense-organ and that of a fully evolved eye, which is able to effect the needful molecular changes in response to a flash as instantaneous as that of lightning.

With regard to luminous stimulation, it is only necessary further to observe that responses were given equally well to direct sunlight, diffused daylight, and to light reflected from a mirror inclined at the polarizing angle. It must also be stated that responses are given to any of the luminous rays of the spectrum when these are employed separately; but that neither the non-luminous rays beyond the red, nor those beyond the violet, appear to exert the smallest degree of stimulating influence.

Electrical Stimulation.

All the excitable parts of all the Medusæ which I have examined are highly sensitive to electrical stimulation, both of the constant and of the induced current.

Exploration with needle-point terminals and induction shocks of graduated strength showed that certain parts or tracts of the nectocalyx are more sensitive than others. The most sensitive parts are those which correspond with the distribution of the main nerve-trunks, *i.e.* round the margin of the nectocalyx and along the course of the radial tubes. The external or convex surface of a nectocalyx or umbrella is totally insensitive to stimulation, and the same statement applies to the whole thickness of the gelatinous substance to which the neuro-muscular sheet is attached.

In all other respects the excitable tissues of the Medusæ in their behaviour towards electrical stimulation conform to the rules which are followed by excitable tissues of other animals. Thus, closure of the constant current acts as a stronger stimulus than does opening of the same, while the reverse is true of the induction shock; and exhaustion super-

venes under the influence of prolonged excitation.
Moreover, I have obtained evidence of that polarization of nerve-tissues under the influence of the constant current, which is known to physiologists by the term "electrotonus;" but it would be somewhat tedious to detail the evidence on this head which I have already published elsewhere.* Tetanus produced by faradaic electricity is not of the nature of an apparently single and prolonged contraction, but that of a number of contractions rapidly succeeding one another, as in the case of the heart under similar excitation. This at least applies to Sarsia. In the case of Aurelia, tolerably strong faradization does cause a more or less well-pronounced tetanus. The continuity of the spasm is, indeed, often interrupted by momentary and partial relaxations. These interruptions are the more frequent the weaker the current; so that, at a certain strength of the latter, the tetanus is of a wild and tumultuous nature; but with strong currents the spasm is tolerably uniform. That in all cases the tetanus is due to summation of contractions may be very prettily shown by the following experiment. An Aurelia is cut into a spiral strip, and all its lithocysts are removed; single induction-shocks are then thrown in with a key at one end of the strip—every shock, of course, giving rise to a contraction wave. If these shocks are thrown in at a somewhat fast rate, two contraction waves may be made at the same time to course along the spiral

* See "Croonian Lectures," 1875. *Philosophical Transactions*, vol. 166, part I. pp. 284-6.

strip, one behind the other; but if the shocks are thrown in at a still faster rate, so as to diminish the distance between any two successive waves, a point soon arrives at which every wave mounts upon its predecessor; and if several waves be thus made to coalesce, the whole strip becomes thrown into a state of persistent contraction.

In this way sustained tetanus, or single contraction waves, or any intermediate phase, may be instantly produced at pleasure. In such experiments, moreover, it is interesting to observe that, no matter how long the strip be, whatever disturbances are set up at one end are faithfully transmitted to the other. For instance, if an Aurelia be cut into the longest possible strip with a remnant of the disk left attached at one end, as represented in Fig. 11 (p. 70), then all the peculiar time relations between successive contractions which are intentionally caused by the experimenter at one end of the strip, are afterwards accurately reproduced at the other end of the strip by the remainder of the disk. Now, as this fact is observable however complex these time relations may be, and however rapidly the successive stimuli are thrown in, I think it is a point of some interest that these complicated relations among rapidly succeeding stimuli do not become blended during their passage along the thirty or forty inches of contractile tissue. The fact, of course, shows that the rate of transmission is so identical in the case of all the stimuli originated, that the sum of the effects of any series of stimuli is delivered at the distal end of the strip, with all its constituent parts as distinct from one another as they were at starting from the proximal end of the strip.

Period of Latency, and Summation of Stimuli.

I shall now give an account of my experiments in the period of latency and the summation of stimuli. To do this, I must first describe the method which I adopted in order to obtain a graphic record of the movements which were given in response to the stimuli supplied. As Aurelia aurita is the only species on which I have experimented in this connection, my remarks under this heading will be confined to it alone.

The method by which I determined the latent period in the case of this species was as follows. A basin containing the Medusa was filled to its brim with sea-water, and placed close beside a smoked cylinder, which, while it lay in a horizontal position, could be rotated at a known rate. The Aurelia* was placed with its concave aspect uppermost, and an inch or two below the surface of the water. The animal was held firmly in this position by means of a pair of compasses thrust through it and forced into a piece of wood, which was fastened to the bottom of the basin. The legs of the com-

---

* It may here be stated that in all the experiments on stimulation subsequently to be detailed, there is no difference to be observed between the behaviour of an entire swimming organ deprived of its ganglia, and that of a portion of any size which may be separated from it.

EXPERIMENTS IN STIMULATION. 51

passes were provided with india-rubber sliders, so that by placing these under the Medusa, the latter might be kept at any elevation in the water which might be desired. The manubrium and lithocysts were now removed, and also a segment of the umbrella. A light straw was then forced through the gelatinous substance of the umbrella in a radial direction, and close to the gap caused by the missing segment. The other, or free, end of this straw was firmly joined to a capillary glass rod, which was suitably bent to avoid contact with the rim of the basin, and also to write on the smoked cylinder. If the straw was not itself sufficient to support the weight of the capillary rod, a small cross-piece of cork might easily be tied to it, so as to add to the flotation power. A part of the excitable tissue was now raised above the surface of the water by means of a disk of cork placed beneath it, and on the part of the tissue thus raised there were placed a pair of platinum electrodes. These electrodes proceeded from an electro-magnetic apparatus, which was arranged in such a way that every time the current in it was opened or closed, it gave an induction shock and moved a lever at the same instant of time. This lever was therefore placed upon the cylinder immediately above the capillary glass-writer which proceeded from the Medusa care being taken to place the two writers in the same line, parallel to the axis of the cylinder. Such being the arrangement, the cylinder was rotated, and thus two parallel lines were marked upon it by the two writers. If the current was now closed, an induction shock was thrown into the tissue at the same instant that the electro-magnet writer recorded the fact, by altering its position on the cylinder. Again, as soon as the paralyzed Medusa responded to the induction shock, the radii of the vacant segment were drawn apart, and in this way a curve was obtained by the other writer on the rotating cylinder. Now, by afterwards dropping a perpendicular line from the point at which the electro-magnet writer changed its position, to the parallel line made by the other writer, and then measuring the distance between the point of contact and the point on the last-mentioned line on which the curve began, the period of latent stimulation was determined. A glance at Figs. 3 and 4 (p. 55) will render this description clear to any one who is not already acquainted with the method, when it is stated that the upper line is a record of the movements of the electro-magnet writer, and the lower line that of the movements of the other writer. It will be observed that the point a in the upper line marks the point at which the induction shock was thrown in; so that by first producing the perpendicular till it meets the lower line at b, and then measuring the distance between the point b and the point c, at which the curve in the lower line first begins, the latent period ($b\,c$) is determined—the time occupied by the rotation of the cylinder from b to c being known.

Summation of Stimuli.—In this way I have been able to ascertain the period of latent stimulation in Aurelia aurita with accuracy. It must be stated at the outset, however, that this period is subject to great variations under certain varying conditions, so that we can only arrive at a just estimation of it by understanding the nature of the modifying causes. To take the simplest cause first, suppose that the paralyzed Aurelia has been left quiet for several minutes in sea-water at forty-five degrees, and that it is then stimulated by means of a single induction shock; the responsive contraction will be comparatively feeble with a very long period of latency, viz. five-eighths of a second. If another shock of the same intensity be thrown in as soon as the tissue has relaxed, a somewhat stronger contraction, with a somewhat shorter latent period, will be given. If the process is again repeated, the response will be still more powerful, with a still shorter period of latency; and so on, perhaps, for eight or ten stages, when the maximum force of contraction of which the tissue is capable will have been attained, while the period of latency will have been reduced to its minimum. This period is three-eighths of a second, or, in some cases, slightly less.

Now, we have here a remarkable series of phenomena, and as it is a series which never fails to occur under the conditions named, I append tracings to give a better idea of the very marked and striking character of the results. The first tracing (Fig. 2) is a record of the successive increments of the responses to successive induction shocks of the same intensity, thrown in at three seconds' intervals—

the cylinder being stationary during each response, and rotated a short distance with the hand during each interval of repose.

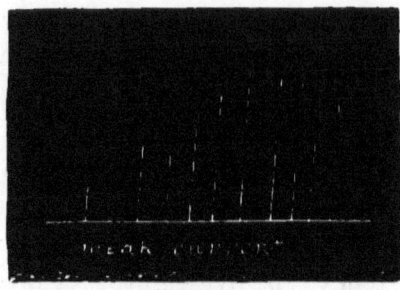

Fig. 2.

The second tracing (Figs. 3 and 4) is a record of the difference between the lengths of the latent period, and also between the strengths of the contraction, in the case (a) of the first of such a series of responses (Fig. 3), and (b) of the last of such a series (Fig. 4). From these tracings it will be manifest, without further comment, how surprising is the effect of a series of stimuli; first, in *arousing* the tissue, as it were, to increased *activity*, and, second, in developing a state of *expectancy*.

In accordance with the now customary terminology, I shall call such a series of responses as are given in Fig. 2 a "staircase." Such a staircase has a greater number of steps in it if caused by a weak current (compare Figs. 2 and 5); and if the strength of the current be suddenly increased after the maximum level of a staircase has been reached by using a feeble current, this level admits of being s'ightly raised (see Fig. 5). Moreover, I find that a

EXPERIMENTS IN STIMULATION. 55

stimulus, which at the bottom of a staircase is of
less than minimal intensity, is able, at the top of a

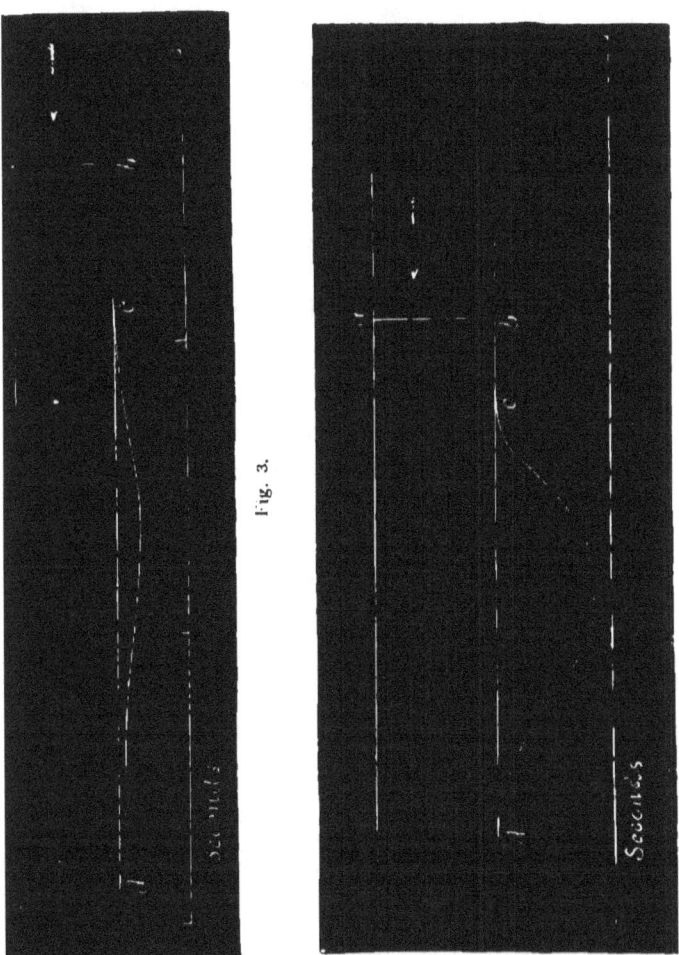

Fig. 3.

staircase, to give rise to a contraction of very nearly
maximum intensity. That is to say, by employing

an induction stimulus of slightly less than minimal intensity in relation to the original irritability of

Fig. 5.

the tissue, no response is given to the first two or three shocks of a series; but at the third or fourth shock a slight response is given, and from that point onward the staircase is built up as usual. This was the case in the experiment of which Fig. 2 is a record, no response having been given to the first two shocks.

With regard to this interesting staircase action, two questions naturally present themselves. In the first place, we are anxious to know whether the arousing effect which is so conspicuous in a staircase series is due to the occurrence of the previous *stimulations*, or to that of previous *contractions*; and, in the next place, we should like to know whether, during the *natural* rhythm of the tissue, each contraction exerts a beneficial influence on its successor, analogous to that which occurs in the case of contractions which are due to *artificial* stimuli. To answer the first of these questions, therefore, I built up a staircase in the ordinary way, and then suddenly transferred the electrodes to the opposite side of the umbrella from that on which they rested while constructing the staircase. On now throwing in another shock at this part of the contractile tissue so remote from the part previously stimulated, the response was a maximum response. Similarly, if the electrodes were transferred in the way just described, not after the maximum effect had been attained, but at any point during the process of constructing a staircase, the response given to the next shock was of an intensity to make it rank as the next step in the staircase. Hence, shifting the position of the electrodes in no wise modifies the peculiar effect we are considering; and this fact conclusively proves that the effect is a general one, pervading the whole mass of the contractile tissue, and not confined to the locality which is the immediate seat of stimulation. Nevertheless, this fact does not tend to prove that the staircase effect depends on the process of contraction as distinguished from the process of stimulation, because the wave of the former process must always precede that of the latter. But, on the other hand, in this connection it is of the first importance to remember the fact already stated, viz. that a current which at the beginning of a series of stimulations is of slightly less than minimal intensity presently becomes minimal, and eventually of much more than minimal intensity—a staircase being thus built up of which the first observable step (or contraction) only occurs in response to the second, third, or even fourth shock of the series. This fact conclusively proves that the staircase effect, at any rate at its commencement, depends on the process of stimulation as distinguished from that of contraction; for it is obvious that the latter process cannot play any part in thus constructing what we may term the invisible steps of a staircase.

To answer the second of the above questions, I placed an Aurelia with its concave surface uppermost, and removed seven of its lithocysts; I then observed the spontaneous discharge of the remaining one, and found it to be conspicuous enough that, after the occurrence of one of the natural pauses (if this were of sufficient duration), the first contraction was feeble, the next stronger, the next still stronger, and so on, till the maximum was attained. This natural staircase action admits of being very prettily shown in another way. If a tolerably large Aurelia is cut into a spiral strip of small width and great length, and if all the lithocysts are removed except one at one end of the strip, it may be observed that, after the occurrence of a natural pause, the first discharge only penetrates perhaps about a quarter of the length of the strip, the next discharge penetrates a little further, the next further, and so on, till finally the contraction waves pass from end to end. On now removing the ganglion, waiting a few minutes, and then stimulating with successive induction shocks, the same progressive penetration is observable as that which previously took place with the ganglionic stimulation. Lastly, the identity of natural and artificial staircase action may be placed beyond all doubt by an experiment in which the effects of induction shocks and of ganglionic discharges are combined. To accomplish this, all the lithocysts save one are removed, and a staircase is then built up in the ordinary way by successive induction shocks. It will now occasionally happen that the ganglion originates a discharge during the process of constructing the staircase, which is being built up by the artificial stimuli; when this happens the resulting contraction takes its proper rank in the series, and this at whatever point the natural contraction happens to come in.

Thus, then, to summarize and conclude these observations, we have seen that if a single stimulation, whether of a natural or artificial kind, is supplied to the excitable tissues of a jelly-fish, a short period, called the period of latency, will elapse, and then the jelly-fish will give a single weak contraction. If, as soon as the tissue has relaxed, the stimulation is again repeated, the period of latency will be somewhat shorter, and will be followed by a somewhat stronger contraction. Similarly, if the stimulation is repeated a third time, the period of latency will be still shorter, and the ensuing contraction still stronger. And so on up to nine or ten times, when the period of latency will be reduced to its *minimum*, while the force of the contraction will be raised to its *maximum;* so that in the jelly-fish, the effect of a series of excitations supplied at short intervals from one another is that of both arousing the tissue into a state of increased *activity*, and also of producing in it a state of greater *expectancy*. We have, moreover, seen that this effect depends upon the repetition of the process of stimulation, and not upon that of the process of contraction.

Now, effects very similar to these have been found to occur in the case of the excitable plants by Dr. Burdon-Sanderson; in the case of the frog's heart by Dr. Bowditch; and in the case of reflex action of the spinal cord by Dr. Stirling. Indeed, the only difference in this respect between these four tissues, so widely separated from one another in the biological scale, consists in the *time* which may be allowed to elapse between the occurrence of the successive stimuli, in order to produce this so-called summating effect of one stimulus upon its successor: the *memory*, so to speak, of the heart-tissue for the occurrence of a former stimulus being longer than the memory of the jelly-fish tissue; while the memory of the latter is longer than that of the plant tissue. And I may here add that even in our own organization we may often observe the action of this principle of the summation of stimuli. For instance, we can tolerate for a time the irritation caused by a crumb in the larynx, but very rapidly the sense of irritation accumulates to a point at which it becomes impossible to avoid coughing. And similarly with tickling generally, the convulsive reflex movements to which it gives rise become more and more incontrollable the longer the stimulation is continued, until they reach a *maximum* point, where, in persons susceptible to this kind of stimulation, the muscular action passes completely beyond the power of the will. Lastly, I may further observe, what I do not think has ever been observed before, that even in the domain of psychology the action of this principle admits of being clearly traced. We find it, for instance, in the rhythmical waves of emotion characteristic of grief, and at the other extreme we find it in the case of the ludicrous. We can endure for a short time, without giving any visible response, the psychological stimulation which is supplied by a comical spectacle; but if the latter continues sufficiently long in a sufficiently ludicrous manner, our appropriate emotion rapidly runs up to a point at which it becomes uncontrollable, and we burst into an explosion of ill-timed laughter. But in this case of psychological tickling, as in the previous case of physiological tickling, some persons are much more susceptible than others. Nevertheless, there can be no doubt that from the excitable tissues of a plant, through those of a jelly-fish and a frog, up even to the most complex of our psychological processes, we have in this recently discovered principle of the summation of stimuli a very remarkable uniformity of occurrence.

Effects of Temperature on Excitability.

I shall now conclude this chapter with a brief statement of the effects of temperature on the excitability of the Medusæ; and before stating my results, I may observe that in all my experiments in this connection I changed the temperature of the

Medusæ by drawing off the water in which they floated with a siphon, while at the same time I substituted water of a different temperature from that which I thus abstracted. In this way, without modifying any of the other conditions to which the animals were exposed, I was able to observe the effects of changing the temperature alone.

With regard to the effect of temperature on the latent period of stimulation, the following table, setting forth the results of one among several experiments, explains itself.

Period of latent stimulation of the deganglionated tissues of Aurelia aurita as affected by temperature:—

Temperature of water (Fahr.).	Period of latent stimulation.
70°	⅕ second
50°	⅓ second
35°	⅔ second
20°	½ second

In the case of each observation, several shocks were administered before the latent period was taken, in order to decrease this period to its *minimum* by the staircase action. When this is not done, the latent period at 20° may be as long as $1\frac{1}{8}$ seconds; but soon after this irritability disappears.

The extraordinary sluggishness of the latent period at very low temperatures is fully equalled by the no less extraordinary sluggishness of the contraction.

In order to render apparent the degree in which both these effects are produced, I here append a pair of tracings which were procured from the same piece of tissue when exposed to the different temperatures named. (N.B.—The seconds are wrongly

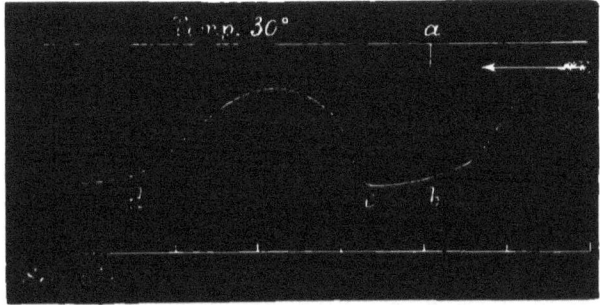

Fig. 6.

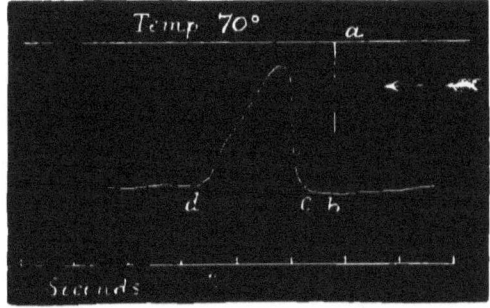

Fig. 7.

marked in Fig. 7; they ought to be the same as in Fig. 6.)

I may as well state here that in water at all temperatures, within the limits where responses to stimuli are given at all, the staircase action admits of being equally well produced; but in order to procure the *maximum* effect for any given temperature, the rate at which the successive stimuli are thrown in must be quicker in warm than in cold water.

CHAPTER IV.

EXPERIMENTS IN SECTION OF COVERED-EYED MEDUSÆ

Amount of Section which the Neuro-muscular Tissues of the Medusæ will endure without suffering Loss of their Physiological Continuity.

THE extent to which the neuro-muscular tissues of the Medusæ may be mutilated without undergoing destruction of their physiological continuity is in the highest degree astonishing. For instance, to begin with the covered-eyed Medusæ, I shall briefly state three modes of section, the results of which serve to show in a striking manner the fact in question.

The annexed woodcuts represent the umbrella of Aurelia aurita, with its manubrium cut off at the base, and the under or concave surface of the umbrella exposed to view, shewing in the centre the ovaries, and radiating from them the branched system of nutrient tubes. The umbrella when fully expanded, as here represented, is about the size of a soup plate, and, as previously stated, all the marginal ganglia are aggregated in the eight marginal bodies or lithocysts. Therefore if the reader will imagine the first of the diagrams (Fig. 8) to be overspread with a disc of muslin, the fibres and mesh of which are finer than those of the finest and closest cobweb, and if he will imagine the mesh of these fibres to start from these marginal ganglia,

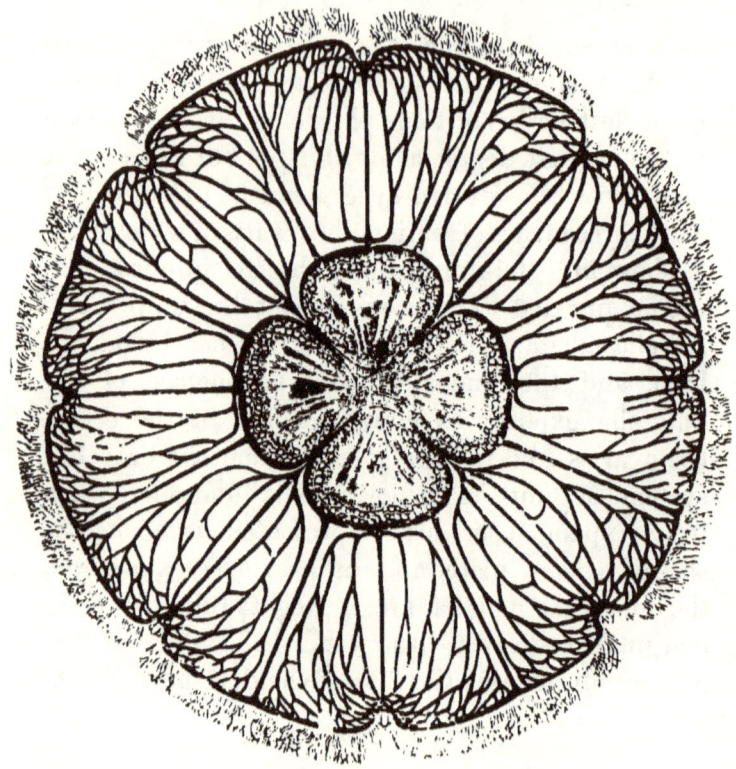

Fig. 8.

he will gain a tolerably correct idea of the lowest nervous system in the animal kingdom. Now, suppose that seven of these eight ganglia are cut out, the remaining one then continues to supply its rhythmical discharges to the muscular sheet of the

SECTION OF COVERED-EYED MEDUSÆ. 67

bell, the result being, at each discharge, two contraction waves, which start at the same instant, one on each side of the ganglion, and which then course with equal rapidity in opposite directions,

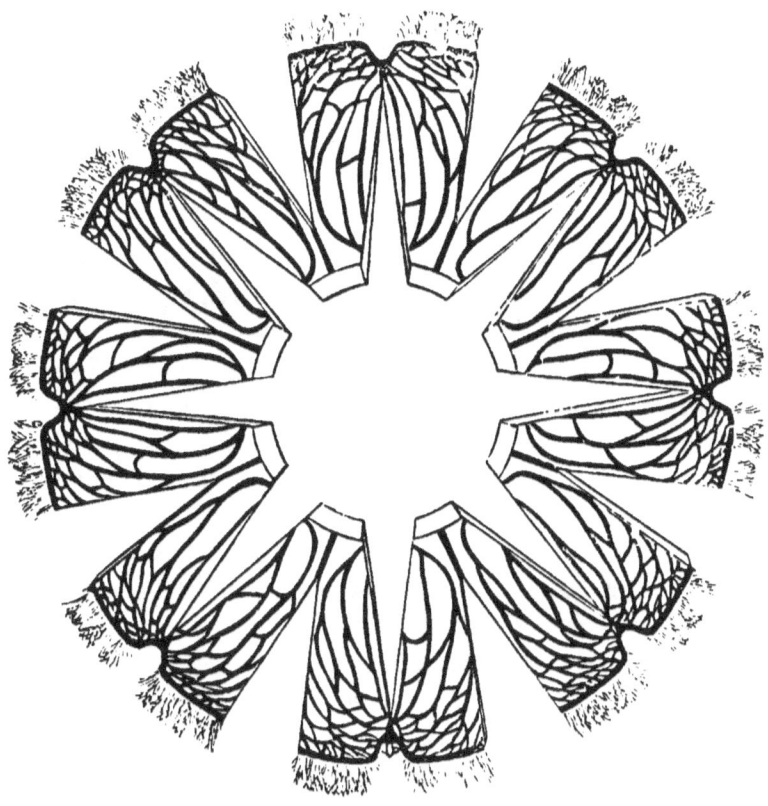

Fig. 9.

and so meet at the point of the disc which is opposite to the ganglion. Suppose, now, a number of radial cuts are made in the disc according to such a plan as this (Fig. 9), wherein every radial cut deeply overlaps those on either side of it. The contraction waves which now originate from the ganglion must either become blocked and cease to pass round the disc, or they must zigzag round and round the tops of these overlapping cuts. Now, remembering that the passage of these contraction waves is presumably dependent on the nervous network progressively distributing the ganglionic impulse to the muscular fibres, surely we should expect that two or three overlapping cuts, by completely severing all the nerve-fibres lying between them, ought to destroy the functional continuity of these fibres, and so to block the passage of the contraction wave. Yet this is not the case; for even in a specimen of Aurelia so severely cut as the one here represented, the contraction waves, starting from the ganglion, continued to zigzag round and round the entire series of sections.

The second mode of section to which I have alluded is as follows (Fig. 10). The central circle (x) stands for an open space cut out of the umbrella; the outer circle indicates the margin of the animal, with all lithocysts save one (l) removed; and the median circular line represents a cut. It will be seen that the effect of this cut is almost completely to sever the mass of tissue at z from the rest of the umbrella, the only connection between them being the narrow neck of tissue at z. Yet, in the case to which I refer, the contraction waves emanating from l passed in the directions represented by the arrows without undergoing any appreciable loss of vigour. Upon completing the circular cut at z, the ring of tissue ($y\ z$) became totally paralyzed, while the outer circle, of course, continued its contractions as before. Now, the neck of tissue at *z* measured only one-eighth of an inch across, while the ring of tissue (*y z*), when cut through and straightened out

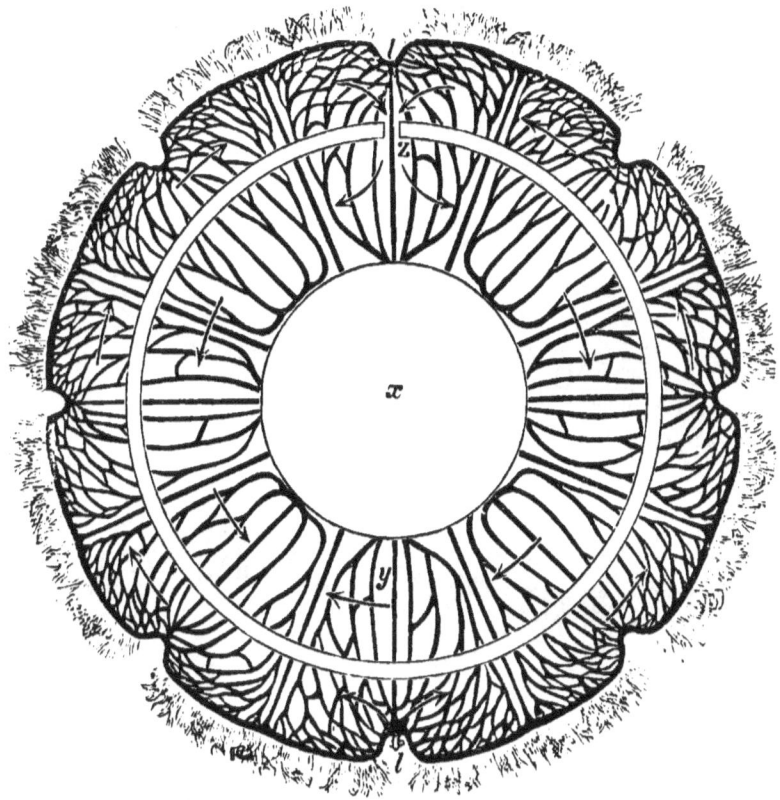

Fig. 10.

upon the table, measured one inch across and sixteen inches in length; that is to say, sixteen square inches of tissue derived its impulse to vigorous contractions through a channel one-eighth of an inch wide,

70   JELLY-FISH, STAR-FISH, AND SEA-URCHINS.

notwithstanding that the latter was situated at the furthest point of the circle from the discharging

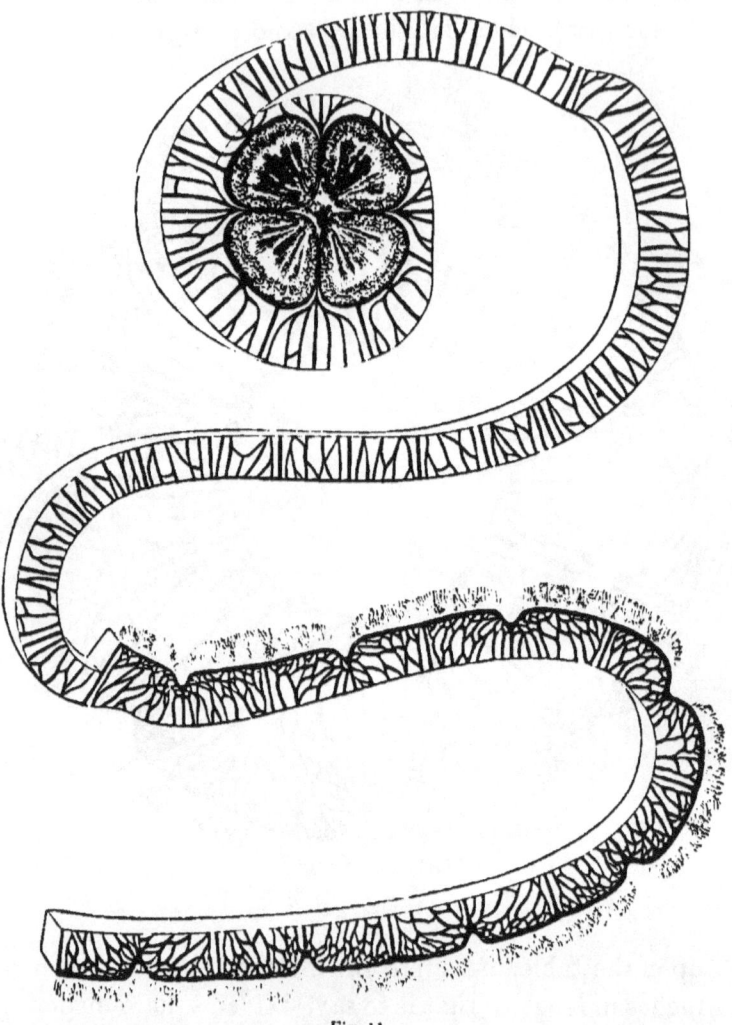

Fig. 11.

lithocyst which the form of the section rendered possible.

Lastly, the third mode of section is represented in the next cut. Here seven of the marginal ganglia having been removed as before, the eighth one was made the point of origin of a circumferential section, which was then carried round and round the bell in the form of a continuous spiral—the result, of course, being this long ribbon-shaped strip of tissue with the ganglion at one end and the remainder of the swimming-bell at the other. Well, as before, the contraction-waves always originated at the ganglion; but now they had to course all the way along the strip until they arrived at its other extremity; and, as each wave arrived at that extremity, it delivered its influence into the remainder of the swimming-bell, which thereupon contracted. Now, in this experiment, when the spiral strip is only made about half an inch broad, it may be made more than a yard long before all the bell is used up in making the strip; and as nothing can well be imagined as more destructive of the continuity of a nerve-plexus than this spiral mode of section must be, we cannot but regard it as a very remarkable fact that the nerve-plexus should still continue to discharge its function. Indeed, so remarkable does this fact appear, that to avoid accepting it we may well feel inclined to resort to another hypothesis, namely, that these contraction-waves do not depend for their passage on the nervous network at all, but that they are of the nature of the muscle waves, or of the waves which we see in undifferentiated protoplasm, where all parts of the mass being equally excitable and equally contractile, however severely we cut the mass, as long as we do not actually divide it, contraction-waves will pass throughout the whole mass. But this very reasonable hypothesis of the contraction-waves in the Medusæ being possibly nothing more than muscle-waves is negatived by other facts, which I shall now proceed to state.

In the first place, if a number of experiments be tried in any of the three modes of section above described, it will be found that extreme variations are manifested as regards the degree of tolerance. In the spiral mode of section, for instance, it will sometimes happen that the contraction-wave will become blocked when the contractile strip is only an inch long, while in other cases (as in the one represented) the wave will continue to pass through a strip more than thirty times that length; and between these two extremes there are all possible grades of tolerance. Now it seems to me that if the tissue through which these contraction-waves pass is supposed (as far as they are concerned) to be of a functionally homogeneous nature, no reason can be assigned why there should be such great differences in the endurance of the tissue in different individual cases; while, if we suppose that the passage of the contraction-waves is more or less dependent on the functional activity of the nervous plexus which we know from microscopical examination to be present, we encounter no such difficulty; for it is almost to be expected that in some cases it would happen that important nerves would soon be encountered by the section, while in other cases it would happen that such nerves would escape the section for a longer distance. It is indeed incredible that any one nerve should happen to pursue a spiral course twice or thrice round the umbrella, and at the same time happen to be concentric with the course pursued by the section; but, as we shall presently see, such an hypothesis as this is not necessary to account for the facts.

Again, in the second place, strong evidence that the passage of the contraction-waves is dependent on the functional activity of the nervous plexus, and therefore that they are not merely muscle-waves, is furnished by the fact that at whatever point in a spiral strip which is being progressively elongated by section the contraction-wave becomes blocked, the blocking is sure to take place *completely* and *exclusively* at that point. Now, as I have tried this experiment a great number of times, and always tried it by carefully feeling the way round (*i.e.* only making a very short continuation of the cut after the occurrence of each contraction-wave, and so very precisely localizing the spot at which the contraction-wave ceased to pass), I can scarcely doubt that in every case the blocking is caused by the cutting through of nerves.*

---

* In a highly interesting paper recently published by Dr. W. H. Gaskell, F.R.S., on "The Innervation of the Heart" (*Journ. of Physiol.*, vol. iv. p. 43, *et seq.*), it is shown that the experiments in section thus far described yield strikingly similar results when performed upon the heart of the tortoise and the heart of the skate. Dr. Gaskell inclines to the belief that in these cases the contraction-waves are merely muscle-waves. There is one important fact, however, which even here seems to me to indicate

But, lastly, the strongest evidence in favour of this view as afforded by the following observations. At the beginning of this treatise I stated that the distinguishing function of nerve consists in its power of conducting stimuli to a distance, irrespective of the passage of a contraction-wave; and I may here add that when a stimulus so conducted reaches a ganglion, or nerve-centre, it causes the ganglion to discharge by so-called "reflex action." Now, this distinguishing function of nerve can plainly be proved to be present in the Medusæ. For instance, take such a section of Aurelia as this one (Fig. 12),

that the propagation of the wave is at least in some measure dependent on nervous conduction. This fact is, that after a contraction-wave has been blocked by the severity of a spiral or other form of section, it may again be made to force a passage under the influence of vagus stimulation.

Moreover, in a paper still more recently published by Drs. Brunton and Cash on "Electrical Stimulation of the Frog's Heart" (*Proc. Roy. Soc.*, vol. xxxv., No. 227, p. 455, *et seq.*) it is remarked, "Another interesting consideration is, whether the stimulus which each cavity of the heart transmits to the succeeding one consists in the propagation of an actual muscular wave, or in the propagation of an impulse along the nerves. The observations of Gaskell have given very great importance to the muscular wave occurring in each cavity of the heart of cold-blooded animals as a stimulus to the contraction of the next succeeding cavity. Our observations appear to us to show that, while this is an important factor, it is not the only one in the transmission of stimuli. . . . We consider that stimuli are also propagated from one chamber of the heart to another through nervous channels: thus we find that irritation of the venus sinus will sometimes produce simultaneous contractions of the auricle and ventricle, instead of the ventricular beat succeeding the auricular in the ordinary way. This we think is hardly consistent with the hypothesis, that a stimulus consists of the propagation of a muscular wave only from the auricle to the ventricle."

SECTION OF COVERED-EYED MEDUSÆ.

wherein the bell has been cut into the form of a continuous parallelogram of tissue with the ovaries and a single remaining ganglion at one end. (The

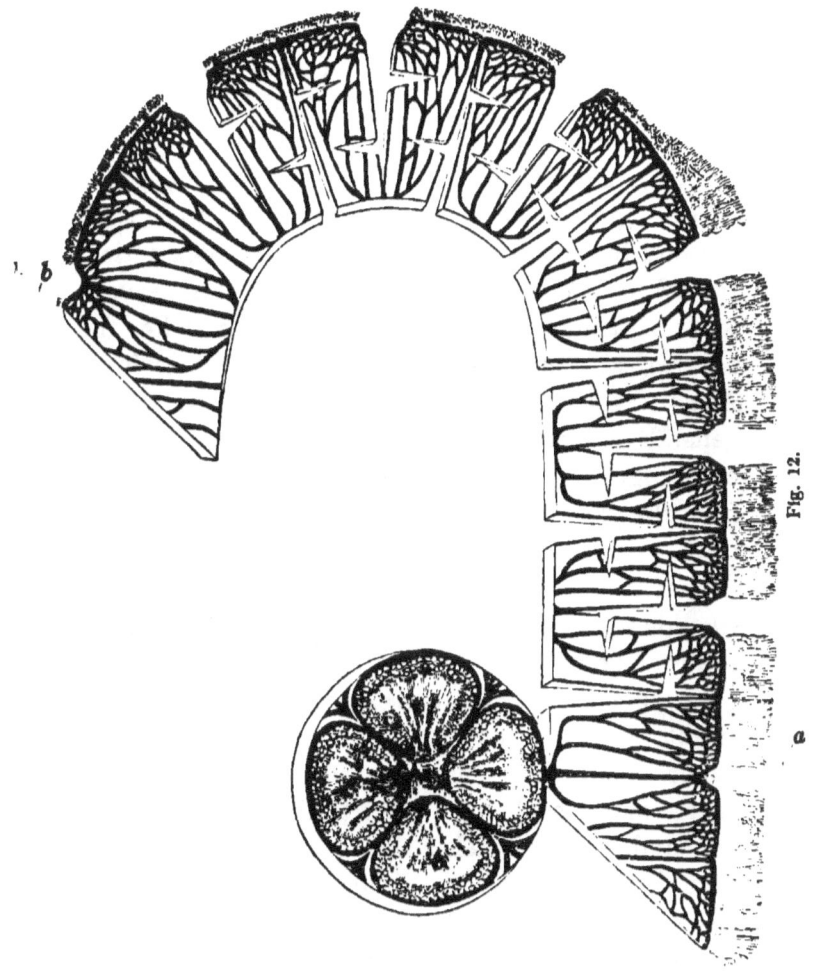

Fig. 12.

cuts interposed in the parallelogram may, for the present, be disregarded.) Now, if the end marked *a* of the neuro-muscular sheet most remote from the ganglion be gently brushed with a camel's hair brush—*i.e.* too gently to start a responsive contraction-wave—the ganglion at the other end will shortly afterwards discharge, as shown by its starting a contraction-wave at its own end of the parallelogram, *b*; thus proving that the stimulus caused by brushing the tissue at the other end, *a*, must have been conducted all the way along the parallelogram to the terminal ganglion, *b*, so causing the terminal ganglion to discharge by reflex action. Indeed, in many cases, the passage of this nervous wave of stimulation admits of being *seen*. For the numberless tentacles which fringe the margin of Aurelia are more highly excitable than is the general contractile tissue of the bell; so that on brushing the end *a* of the parallelogram remote from the ganglion, the tentacles at this end respond to the stimulus by a contraction, then those next in the series do the same, and so on—a wave of contraction being thus set up in the tentacular fringe, the passage of which is determined by the passage of the nervous wave of stimulation in the superjacent nervous network. This tentacular wave is in the illustration represented as having traversed nearly half the whole distance to the terminal ganglion, and when it reaches that ganglion it will cause it to discharge by reflex action, so giving rise to a visible wave of muscular contraction passing in the direction *b a*, opposite to that which the nervous or tentacular wave had previously pursued. Now this tentacular wave, being an optical expression of a passage of a wave of stimulation, is a sight as beautiful as it is unique; and it affords a first-rate opportunity of settling this all-important question, namely, Will this conductile or nervous function prove itself as tolerant towards a section of the tissue as the contractile or muscular function has already proved itself to be? For, if so, we shall gain nothing on the side of simplicity by assuming that the *contraction*-waves are merely muscle-waves, so long as the *conduction* or *undoubtedly nervous* waves are equally able to pass round sections interposed in their path. Briefly, then, I find that the nervous waves of stimulation are quite as able to pass round these interposed sections as are the waves of contraction. Thus, for instance, in this specimen (Fig. 12), the tentacular wave of stimulation continued to pass as before, even after I had submitted the parallelogram of tissue to the tremendously severe form of section which is represented in the illustration; and this fact, in my opinion, is one of the most important that has been brought to light in the whole range of invertebrate physiology. For what does it prove? It proves that the distinguishing function of nerve, where it first appears upon the scene of life, admits of being performed vicariously to almost any extent by all parts of the same tissue-mass. If we revert to our old illustration of the muslin as representing the nerve-plexus, it is clear that, however much we choose to cut the sheet of muslin with such radial or spiral sections as are represented in the illustrations, one could always trace the threads of the muslin with a needle round and round the disc, without once interrupting the continuity of the tracing; for on coming to the end of a divided thread, one could always double back on it and choose another thread which might be running in the required direction. And this is what we are now compelled to believe takes place in the fibres of this nervous network, if we assume that these visible fibres are the only conductile elements which are present. Whenever a stimulus wave reaches a cut, we must conclude that it doubles back and passes into the neighbouring fibres, and so on, time after time, till it succeeds in passing round and round any number of overlapping cuts.

This is, no doubt, as I have already observed, a very remarkable fact; but it becomes still more so when we have regard to the histological researches of Professor Schäfer on the structural character of this nerve-plexus. For these researches have shown that the nerve-fibres which so thickly overspread the muscular sheet of Aurelia do not constitute a true plexus, but that each fibre is comparatively short and nowhere joins with any of the other fibres; that is to say, although the constituent fibres of the network cross and recross one another in all directions—sometimes, indeed, twisting round one another like the strands of a rope—they can never be actually seen to join, but remain anatomically insulated throughout their length. So that the simile by which I have represented this nervous network—the simile, namely, of a sheet of muslin overspreading the whole of the muscular sheet—is, as a simile, even more accurate than has hitherto appeared; for just as in a piece of muslin the constituent threads, although frequently meeting one another, never actually coalesce, so in the nervous network of Aurelia, the constituent fibres, although frequently in contact, never actually unite.

Now, if it is a remarkable fact that in a fully differentiated nervous network the constituent fibres are not improbably capable of vicarious action to almost any extent, much more remarkable does this fact become when we find that no two of these constituent nerve-fibres are histologically continuous with one another. Indeed, it seems to me we have here a fact as startling as it is novel. There can scarcely be any doubt that *some* influence is communicated from a stimulated fibre a to the adjacent fibre b at the point where these fibres come into close apposition. But what the nature of the process may be whereby a disturbance in the excitable protoplasm of a sets up a sympathetic disturbance in the anatomically separate protoplasm of b, supposing it to be really such—this is a question concerning which it would as yet be premature to speculate. But I think it may be well for physiologists to keep awake to the fact that a process of this kind probably takes place in the case of these nerve-fibres. For it thus becomes a possibility which ought not to be overlooked, that in the fibres of the spinal cord, and in ganglia generally, where histologists have hitherto been unable to trace any anatomical or structural continuity between cells and fibres, which must nevertheless be supposed to possess physiological or functional continuity—it thus becomes a possibility that in these cases no such anatomical continuity exists, but that the physiological continuity is maintained by some such process of physiological induction as probably takes place among the nerve-fibres of Aurelia.*

I have now to detail another fact of a very puzzling nature, but one which is certainly of importance. When the spiral section is performed on Aurelia aurita, and when, as a consequence, the contraction-waves which traverse the elongating strip become at some point suddenly blocked, if the section be stopped at this point it not unfrequently happens that after a time the blocking suddenly ceases, the contraction-waves again passing from the strip into the umbrella as freely as they did before the section reached the point at which the blocking occurred. The time required for this restoration of physiological continuity is very variable, the limits being from a few seconds to an hour or more; usually, however, it is from two to four minutes. This process of re-establishing the physiological connections, although rapid, is not so instantaneous as is that of their destruction by section. In general it requires the passage of several contraction-waves before the barrier to the passage of succeeding waves is completely thrown

---

* That it can scarcely be *electrical induction* would seem to be shown by the fact that such effects can only be produced on nerves by strong currents, and also by the fact that the saline tissues of the swimming-bell must short-circuit any feeble electrical currents as soon as they are generated.

down. The first wave which effects a passage appears to have nearly all its force expended in overcoming the barrier, the residue being only sufficient to cause a very feeble, and sometimes almost imperceptible, contraction of the umbrella. The next wave, however, passes across the barrier with more facility, so that the resulting contraction of the umbrella is more decided. The third wave, again, causes a still more pronounced contraction of the umbrella; and so on with all succeeding waves, until every trace of the previous blocking has disappeared. When this is the case, it generally happens that the strip will again admit of being elongated for a short distance before a blocking of the contraction-waves again supervenes. Sometimes it will be found that this second blockage will also be overcome, and that the strip will then admit of being still further elongated without the passage of the waves being obstructed; and so on occasionally for three or four stages.

The same series of phenomena may be shown in another way. If a contractile strip of tolerable length be obtained, with the waves passing freely from one end to the other, and if a series of parallel and equidistant cuts be made along one side of the strip, in a direction at right angles to the length, and each cut extending two-thirds of the breadth of the strip, the chances are in favour of the contraction-waves being wholly unaffected by the sections, however numerous these may be. But now, if another series of parallel and equidistant cuts of the same length as the first ones, and alternating with them, be made along the other side of the contractile strip, the result is, of course, a number of interdigitating cuts ; and it is easy to see that by beginning with a few such cuts and progressively increasing their number, a point must somewhere be reached at which one portion will become physiologically separated from the rest. The amount of such section, however, which contractile strips will sometimes endure is truly surprising. I have seen such a strip twenty inches long by one and a half inches wide with ten such cuts along each side, and the contraction-waves passing without impediment from end to end. But what I wish more especially to observe just now is, that by progressively increasing the number of such interdigitating cuts up to the point at which the contraction-wave is blocked, and then leaving the tissue to recover itself, in many cases it will be observed that the blocking is sooner or later overcome; that on then adding more interdigitating cuts the blocking again supervenes; but that in time it may again be overcome, and so on. It is, however, comparatively rare to find cases in which blocking is overcome twice or thrice in succession.

Section is not the only way in which blocking of waves may be caused in contractile strips. I find that pressure, even though very gentle, exerted on any part of a strip causes a blocking of the waves at that part, even after the pressure has been removed. If the pressure has been long continued, after its removal the blocking will probably be permanent; but if the pressure has been only of

SECTION OF COVERED-EYED MEDUSÆ. 83

short duration, the blocking will most likely be transitory. Even the slight strains caused by handling a contractile strip in the air are generally followed by a decrease in the rate of the waves, and sometimes by their being completely blocked. Other methods by which the passage of waves in contractile strips admits of being blocked will be alluded to farther on.

Now, in all these cases of temporary blocking we must conclude that when the contraction-waves succeed in at last forcing a passage, some structural change has taken place in the tissue at the region of injury, corresponding with the functional change of the re-establishment of physiological continuity. The waves previously stopped at a certain point of section or otherwise, after beating for a time on the physiological barrier, are at last able to throw down the barrier, and thenceforward to proceed on their way unhindered. What, then, is the nature of the structural change which has taken place?

In the early days of this research, before the presence of a nerve-plexus had been proved histologically, I argued in favour of such a plexus on the grounds furnished by many of the foregoing experiments; and at a lecture given in the Royal Institution I ventured to say that if a careful investigation of the histology of these tissues should fail to show the plexus which the result of those experiments required me to assume, we should still be compelled to suppose that the plexus was present, although not sufficiently differentiated to admit of being seen. I further ventured to suggest that in this event the facts just stated might be taken to substantiate the theory of Mr. Herbert Spencer on the genesis of nerve-tissue in general. This theory is that which supposes incipient conductile tissues, or rudimentary nerve-fibres, to be differentiated from the surrounding contractile tissues, or homogeneous protoplasm, by a process of integration which is due simply to use; so that just as water continually widens and deepens the channel through which it flows, so molecular or nervous waves of stimulation, by always flowing through the same tissue-tracts, tend ever more and more to excavate for themselves functionally differentiated lines of passage.

Such being Mr. Spencer's theory, I applied it hypothetically to the above facts in the words which I may here quote.

"As the successive waves beat rhythmically on the area of obstruction, more or less of the molecular disturbances must every time be equalized through these lines of discharge, which from the first have been almost sufficient to maintain the physiological continuity of the tissue. Therefore, according to the hypothesis, every wave that is blocked imposes upon these particular lines of discharge a much higher degree of functional activity than they were ever before required to exercise; and this greater activity causing in its turn greater permeability, a point will sooner or later arrive at which these lines of discharge, from having been *almost*, become *quite* able to draft off sufficient molecular motion, or stimulating influence, to carry on the contraction-waves beyond the areas of previous blocking. In such instances, of course, we should expect to find what I always observed to be the case, viz. that the first contraction-wave which passes the barrier is only very feeble, the next stronger, the next still stronger, and so on, according as the new passage becomes more and more permeable by use, until at last the contraction-waves pour over the original barrier without any perceptible diminution of their force. In some cases, by exploring with graduated stimuli and needle-point terminals, I was able to ascertain the precise line through which this eruption of stimulating influence had taken place."

I have now to state the effect upon this hypothesis which in my opinion has been produced by the histological proof that the plexus in question is composed of fully differentiated nerves. Briefly, then, I think that the hypothesis still holds to the extent of being the only one available whereby to explain the facts; but there is this great difference, viz. that the hypothesis need not now be applied to the genesis of nerve-tissue out of comparatively undifferentiated contractile tissue, but rather to the increasing of the functional activity of already well-differentiated nerve-tissue. In other words, we have not now to suppose that nerve-tissue is formed *de novo* in the region of blocking; but, in my opinion, we still have to suppose that the nerve-fibres which were already there have their functional capabilities so far improved by the greater demand imposed upon them, that whereas at first

they were not able, eventually they became able to
draft off enough molecular disturbance to carry on
a stimulus adequate to cause a muscular contraction.
It will be observed that the difference thus ex-
pressed is one of considerable importance; for now
the facts cease to lend any countenance to Mr.
Spencer's theory touching the formation of nerves
out of protoplasm, or other contractile material.
They continue, however, to countenance his views
touching the improvement of functional capacity
which nerve-fibres, when already formed, undergo
by use; and this, which is in itself an important
matter, is the point with which I was mainly concerned in the lecture of the Royal Institution just
alluded to. For, as I then observed, in this theory
of nerve-fibres becoming more and more functionally
developed by use, we probably have a physical explanation, which is as full and complete as such an
explanation can ever be, of the genesis of mind.
"For from the time that intelligence first dawned
upon the scene of life, whenever a new relation had
to be established in the region of mind, it could only
be so established in virtue of some new line of discharge being excavated through the substance of
the brain. The more often this relation had to
be repeated in the mind, the more often would
this discharge require to take place in the brain,
and so the more easy would every repetition of
the process become. . . . Thus it is, according to
the theory, that there is always a precise proportion
between the constancy with which any relations
have been joined together during the history of

intelligence, and the difficulty which intelligence now experiences in trying to conceive of such relations as disjoined. Thus it is that, even during the history of an individual intelligence, 'practice makes perfect,' by frequently repeating the needful stimulations along the same lines of cerebral discharge, so rendering the latter ever more and more permeable by use. Thus it is that a child learns its lessons by frequently repeating them; and thus it is that all our knowledge is accumulated."*

Rate of Transmission of Stimuli.

The rate at which contraction-waves traverse spiral strips of Aurelia is variable. It is largely determined by the length and width of the strip; so

---

* I have associated the above theory of nerve-genesis with the name of Mr. Spencer, because it occupies so prominent a place in his "Principles of Psychology." But from what I have said in the text, I think it is clear that the theory, as presented by Mr. Spencer, consists of two essentially distinct hypotheses—the one relating to the formation of nerve-tissue out of protoplasm, and the other to the increase of functional capacity in a nerve-fibre by use (a third hypothesis of Mr. Spencer relating to the formation of ganglion-tissue does not here concern us). The latter hypothesis, however, ought not to be associated with Mr. Spencer's name without explaining that it has likewise occurred to other writers, the first of which, so far as I can ascertain, was Lamarck, who says, "Dans toute action, le fluide des nerfs qui la provoque, subit un mouvement de déplacement qui y donne lieu. Or, lorsque cette action a été plusieurs fois répétée, il n'est pas douteux que le fluid qui l'a exécutée, ne se soit frayé une route, qui lui devient alors d'autant plus facile à parcourir, qu'il l'a effectivement plus souvant franchie, et qu'il n'ait lui-même une aptitude plus grand à suivre cette route frayée que celles qui le sont moins." ("Phil. Zool.," tom ii. pp. 318-19.)

that the best form of strip to use for the purpose of ascertaining the *maximum* rate is one which I shall call the circular strip. A circular strip is obtained by first cutting out the central bodies (*i.e.* manubrium and ovaries), and then, with a single radial cut, converting the animal from the form of an open ring to that of a continuous band. I distinguish this by the name "circular" band or strip, because the two ends tend to preserve their original relative positions, so giving the strip more or less of a circular form. Such a strip has the advantage of presenting all the contractile tissue of the swimming-bell in one continuous band of the greatest possible width, and is therefore the form of strip that yields the *maximum* rate at which contraction-waves are able to pass. The reason why the *maximum* rate should be the one sought for is because this is the rate which must most nearly approximate the natural rate of contraction-waves in the unmutilated animal. This rate, at the temperature of the sea and with vigorous specimens, I find to be eighteen inches per second.

In a circular strip the rate of the waves is uniform over the whole extent of the strip; so that the time of their transit from one point to another varies directly as the length of the strip. But on now narrowing such a strip, although the rate is thus slowed, the relation between the narrowing and the slowing is not nearly so precise as to admit of our saying that the rate varies inversely as the width. The following figure will serve to show the proportional extent to which the passage of contraction-waves is retarded by narrowing the area through which they pass :—

Time from end to end of a
 circular strip ... ...
Time after width has been
 reduced to one-half ...
Time after width has been
 reduced to one-quarter...
Time after width has been
 reduced to one-eighth ...

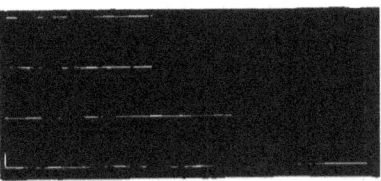

Fig. 13.

In such experiments it generally happens, as here represented, that reducing the width of a circular strip by one-half produces no effect, or only a slight effect, on the rate, while further narrowing to the degree mentioned produces a conspicuous effect. I may also state that if, as occasionally happens, the immediate effect of narrowing a circular strip to one-half is to temporarily block the contraction-waves, when the latter again force their passage, their rate is slower than it was before. It seems as if the more pervious tissue tracts having been destroyed by the section, the less pervious ones, though still able to convey the contraction-wave, are not able to convey it so rapidly as were the more pervious tracts.

In order to ascertain whether certain zones of the circular contractile sheet in all individuals habitually convey more of the contractile influence than do other zones, I tried a number of experiments in the following form of section. Having made a circular strip, I removed all the lithocysts save one, and then cut the strip as represented in Fig. 14. On

90 JELLY-FISH, STAR-FISH, AND SEA-URCHINS.

now stimulating the end a, or on watching the lithocyst there discharge, the resulting contraction-wave would be observed to bifurcate at b, and then

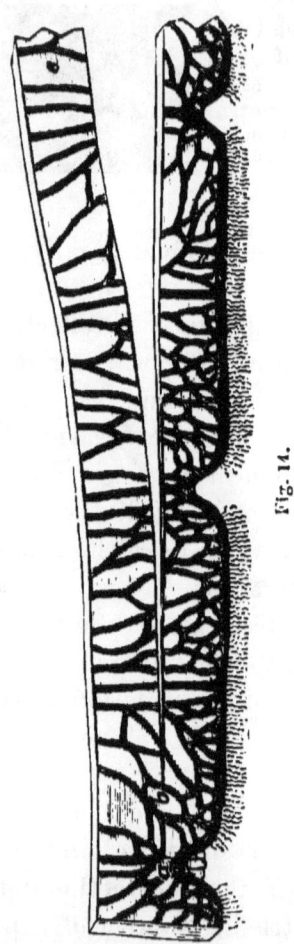

Fig. 14.

pass on as two separate waves through the zones, $b\ c$, $b\ d$. Now, as these two waves were started at the same instant of time, they ran, as it were, a race in the two zones, and in this way the eye could judge with perfect ease which wave occupied the shortest time in reaching its destination. This experiment could be varied by again bisecting each of these two zones, thus making four zones in all, and four waves to run in each race. A number of experiments of this kind showed me that there is no constancy in the relative conductivity of the same zones in different individuals. In some instances, the waves occupy less time in passing through the zone bc than in passing through the zone bd; in other instances, the time in the two zones is equal; and, lastly, the converse of the first-mentioned case is of equally frequent occurrence. Very often the waves become blocked in bc, while they continue to pass in bd, and *vice versâ*. Now, all these various cases are what we might expect to occur, in view of the variable points at which contraction-waves become blocked in spiral strips, etc.; for if the contractile tissues are not functionally homogeneous, and if the relatively pervious conductile tracts are not constant as to their position in different individuals, the results I have just described are the only ones that could be yielded by the experiments in question. Considering, however, that in these experiments the central zones are not so long as the peripheral zones, I think it may fairly be said that the conductile power of the latter is greater than that of the former; for, otherwise, the above experiments ought to yield a large majority of races won by the waves that course through the central zones, and this is not the case. Indeed, it is surprising how often the race is, as it were, neck and neck, thus showing that the relative conductivity of all the zones is precisely adjusted to their relative lengths; and forasmuch as in the unmutilated animal this adjustment must clearly serve the purpose of securing to the contraction-wave a passage of uniform rate over the whole radius of the umbrella, I doubt not that, if it were possible to perform the race-course section without interrupting any of the lines of conduction-tissue, neck and neck races would be of invariable occurrence.

Interdigitating cuts, as might be expected, prolong the time of contraction-waves in their passage through the tissue in which the cuts are interposed. For example, in a spiral strip measuring twenty-six inches in length, the time required for the passage of a contraction-wave from one end to the other is represented by the line $a\,b$ in the annexed woodcut. But after twenty interdigitating cuts had been interposed, ten on each side of the strip, the time increased to $c\,d$, the line $e\,f$ representing one second. And more severe forms of section are, of course, attended with a still more retarding influence.

Fig. 15.

The effects of temperature on the rate of contraction-waves are very striking. For instance, in a rather narrow strip measuring twenty-eight inches long and one and a half inches wide, the following variations in rate occurred:—

SECTION OF COVERED-EYED MEDUSÆ.

Temperature of water.	Time occupied in passage of contractile waves.
26°	4 seconds.
32°	3 seconds.
42°	2⅔ seconds.
65°	2 seconds.
75°	1¾ seconds.
85°	Blocked.

Or, adopting again the graphic method of statement, these variations may be represented as follows:—

Fig. 16.

Submitting a contractile strip to slight strains has also the effect of retarding the rate of the waves while they pass through the portions of the strip which have been submitted to strain. The method of straining which I adopted was to pass my finger below the strip, and then, by raising my hand, to bring a portion of the strip slightly above the level of the water. The irritable or contractile surface was kept uppermost, and therefore suffered a gentle strain; for the weight of the tissue on either side of the finger made the upper surface somewhat convex. By passing the finger all the way along the strip in this way, the latter might be gently strained throughout its entire length, the degree of straining being determined by the height out of the water to which the tissue was raised. Of course, if the strip is too greatly strained, the contraction-waves become blocked altogether, as described above; but shortly before this degree of straining was reached, I could generally observe that the rate of the waves was diminished. To give one instance, a contractile strip measuring twenty-two inches had the rate of its waves taken before and after straining of the kind described. The result was as follows:—

Before straining...
After straining ...
One second ...

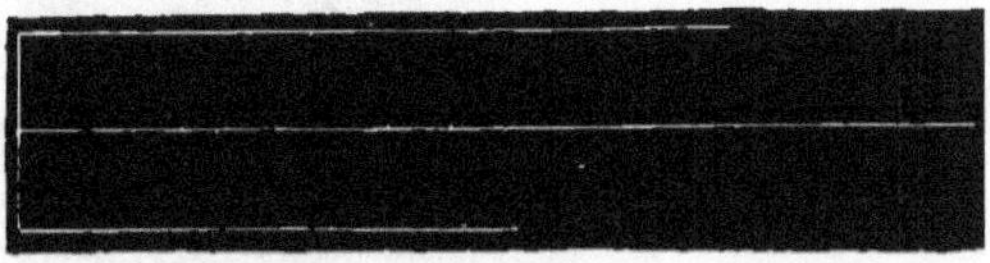

Fig. 17.

Immediately after severe handling of this kind, the retardation of contraction-waves is sometimes even more marked than here represented; but I think this may be partly due to shock, for on giving the tissue a little while to recover, the rate of the waves becomes slightly increased.

Anæsthetics likewise have the effect of slowing the rate of contraction-waves before blocking them. Taking, for instance, the case of chloroform, a narrow spiral strip between one and two feet long was immersed in sea-water containing a large dose of the anæsthetic; the observations being taken at six seconds' intervals, the following were the results:—

SECTION OF COVERED-EYED MEDUSÆ.

Normal water ... ...
Six seconds after trans-
ference to chloroform

Six seconds later...

Six seconds later... ...

Six seconds later...

Six seconds later...

One second

Fig. 18.

In such experiments, the recovery of the normal rate in unpoisoned water is gradual. Taking, for instance, the case of a spiral strip in morphia (Fig. 19), it will be seen that the original rate did not fully return. Some substances, however, exert a more marked permanent effect of this kind than do weak solutions of morphia. Here, for instance, is an experiment with alcohol (see Fig. 20).

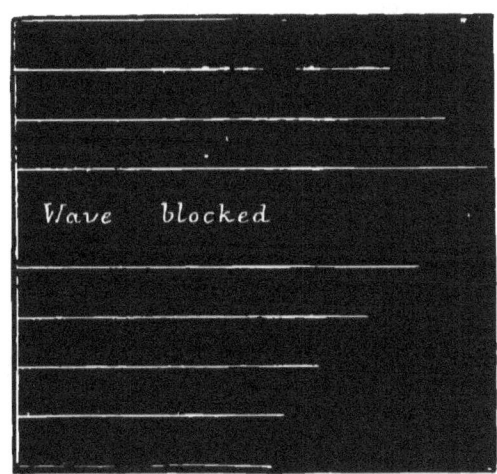

In normal water ...
Quarter of an hour after
exposure to morphia

One minute after strengthening dose ...
Four minutes later, and
just before blocking of
wave ...
Fifteen seconds later, wave
continuing blocked .
Immediately after passage
of wave on return
to normal sea-water ...

Four minutes later

Quarter of an hour later...

An hour later

One second ... ...

Fig. 19.

In normal water ... ...
Quarter of an hour after exposure to weak dose ...
Two minutes after strengthening of dose ... ...
Five minutes later, and just before blocking of wave
Fifteen seconds later, wave continuing blocked ...
Immediately after passage of wave on restoration to normal sea-water ... ...

An hour later ...

One second ...

Wave blocked

Fig. 20.

From these experiments, however, it must not be definitely concluded that it is the anæsthesiating property of such substances which exerts this slowing and blocking influence on contraction-waves, for I find that almost any foreign substance, whether or not an anæsthetic, will do the same. That nitrite of amyl, caffein, etc., should do so, one would not be very surprised to hear; but it might not so readily be expected that strychnine, for instance, should block contraction-waves; yet it does so, even in doses so small as only just to taste bitter. Nay, even fresh water completely blocks contraction-waves after the strip has been exposed to its influence for about half an hour, and exerts a permanently slowing effect after the tissue is restored to sea-water. These facts show the extreme sensitiveness of the neuro-muscular tissues of the Medusæ to any change in the character of their surrounding medium, a sensitiveness which we shall again have occasion to comment upon when treating of the effects of poisons.

In conclusion, I may mention an interesting fact which is probably connected with the summation of stimuli before explained. When a contractile strip is allowed to rest for a minute or more, and when a wave is then made to traverse it, careful observation will show that the passage of the first wave is slower than that of its successor, provided the latter follows the former after not too great an interval of time. The difference, however, is exceedingly slight, so that to render it apparent at all the longest possible strips must be used, and even then the experimenter may fail to detect the difference, unless he has been accustomed to signalling, by which method all these observations on rate have to be made.

Stimulus-waves.

The rate of transmission of tentacular waves is only one-half that of contraction-waves, viz. nine inches a second. This fact appeared to me very remarkable in view of the consideration that the tentacular wave is the optical expression of a stimulus-wave, and that there can be no conceivable use in a stimulus-wave being able to pass through contractile tissue independently of a contraction-wave, unless the former is able to travel more rapidly than the latter; for the only conceivable use of the stimulus-wave is to establish physiological harmony between different parts of the organism, and if this wave cannot travel more rapidly than a contraction-wave which starts from the same point, it would clearly fail to perform this function.

In view of this anomaly, I was led to think that if the rate of the stimulus-wave is dependent in a large degree on the strength of the stimulus that starts it, the slow rate of nine inches a second might be more than doubled, if, instead of using a stimulus so gentle as not to start a contraction-wave, I used a stimulus sufficiently strong to do this. Accordingly I chose a specimen of Aurelia wherein the occurrence of tentacular waves was very conspicuous, and found, as I had hoped, that every time I stimulated too gently to start a contraction-wave, the tentacular wave travelled only at the rate of nine inches a second; whereas, if I stimulated with greater intensity, I could always observe the tentacular wave coursing an inch or two in front of the contraction-wave.

It is remarkable, however, that in this, as in all the other specimens of Aurelia which I experimented upon, the reflex response of the manubrium was equally long, whatever strength of stimulus I applied to the umbrella; or, at any rate, the time was only slightly less when a contraction-wave had passed than when only a tentacular wave had done so. The loss of time, however, appears to take place in the manubrium itself, where the rate of response is astonishingly slow. Thus, if one lobe be irritated, it is usually from four to eight seconds before the other lobes respond. But the time required for such sympathetic response may be even more variable than this—the limits I have observed being as great as from three to ten seconds. In all cases, however, the response, when it does occur, is sudden, as if the distant lobe had then for the first time received the stimulus. Moreover, one lobe—usually one of those adjacent to the lobe directly irritated—responds before the other two, and then a variable time afterwards the latter also respond. This time is, in most cases, comparatively short, the usual limits being from a quarter of a second to two seconds. How much of these enormous intervals is occupied by the period of ganglionic latency, and how much by that of transmission, it is impossible to say; but I have determined that the rate of transmission from the end of a lobe of the manubrium to a lithocyst (deducting a second for the double period of latent stimulation) is the same as the rate of a tentacular wave, viz. nine inches a second. The presumption, therefore, is that the immense lapse of time required for reflex response on the part of the manubrium is required by the lobular ganglia, or whatever element it is that here performs the ganglionic function.

Exhaustion.

In various modes of section of Aurelia I have several times observed a fact that is worth recording. It sometimes happens that when the connecting isthmus between two almost severed areas of excitable tissue is very narrow, the passage of contraction-waves across the isthmus depends upon the freshness, or freedom from exhaustion, of the tissue which constitutes the isthmus. That is to say, on faradizing one of the two tissue-areas which the isthmus serves to connect, the resulting contraction-waves will at first pass freely across the isthmus; but after a time it may happen in some preparations that every now and then a contraction-wave fails to pass across the isthmus. When this is the case, if the stimulation is still continued, a greater and greater proportion of waves fail to pass across the isthmus, until perhaps only one in every five or six becomes propagated from the one area to the other. If single induction-shocks be then substituted for the faradaic stimulation, it may be found that by leaving an interval of four or five seconds between the successive shocks, every wave that is started in the one area will be propagated across the isthmus to the other area. But if the interval between the successive shocks be reduced to two or three seconds, every now and then a wave will fail to pass across the isthmus; and if the interval be still further reduced to one second, or half a second, comparatively few of the waves will pass across. Now, however, if the tissue be allowed five minutes' rest from stimulation, and the single shocks be thrown in at one second's intervals, all the first six or ten waves will pass across the isthmus, after which they begin to become blocked as before. It may be observed also that when the waves are thus blocked, owing to exhaustion of the connecting isthmus, they may again be made to force a passage by increasing the intensity of the stimulation, and so giving rise to stronger waves having a greater power of penetration. Thus, on re-enforcing the electrical stimulus with the simultaneous application of a drop of spirit, the resulting waves of contraction are almost sure to pass across the isthmus, even though this has been exhausted in the manner just described.

Ganglia appearing to assert their Influence at a Distance from their own Seat.

Another fact, which I have several times noticed during my sections of Aurelia, also deserves to be recorded. I have observed it under several modes of section, but it will be only necessary to describe one particular case.

In the Aurelia of a portion of which the accompanying woodcut (p. 102) is a representation, seven of the lithocysts were removed, while the remaining one was almost entirely isolated from the general contractile tissue by the incisions aa, bb, cc. The lithocyst continued to animate the tissue-area $xxxx$, and through the connecting passage y the contraction-waves spread over the remainder of the sub-umbrella tissue $zzzz$. So far, of course, the facts were normal; but very frequently it was observed that the contraction-waves did not start from the lithocyst, or from the area $xxxx$, but from the point o in the area zz. After this origination of the contraction-waves from the point o had been observed a great number of times, I removed the lithocyst. The effect was not only to prevent the further origination of contraction-waves in the area $xxxx$, but also to prevent their further origination from the point o, the entire umbrella thus becoming paralyzed. Hence, before the removal of the lithocyst, the contraction-waves which originated at the point *o*, no less than those which originated at the lithocyst itself, must in some way or other have been due to the ganglionic influence emanating from the lithocyst and asserting itself at the distant point *o*.

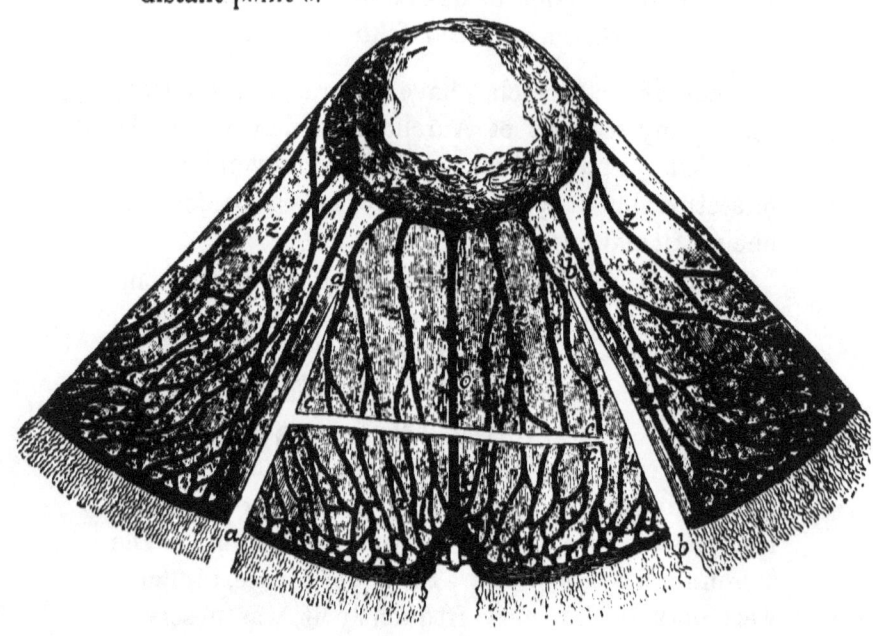

Fig. 21.

This property, which lithocysts sometimes present, of asserting their ganglionic influence at a distance from their own locality, can only, I think, be explained by supposing that at the point where under these circumstances the contractions originate, there are situated some scattered ganglionic cells of considerable functional power, but yet not of power enough to originate contraction-waves unless re-enforced by some stimulating influence, which reaches them from the lithocyst through the nervous plexus.

Regeneration of Tissues.

The only facts which remain to be stated in the present chapter have reference to the astonishing rapidity with which the excitable tissues of the Medusæ regenerate themselves after injury. In this connection I have mainly experimented on Aurelia aurita, and shall, therefore, confine my remarks to this one species.

If with a sharp scalpel an incision be made through the tenuous contractile sheet of the sub-umbrella of Aurelia, in a marvellously short time the injury is repaired. Thus, for instance, if such an incision be carried across the whole diameter of the sub-umbrella, so as entirely to divide the excitable tissues into two parts while the gelatinous tissues are left intact, the result of course is that physiological continuity is destroyed between the one half of the animal and the other, while the form of the whole animal remains unchanged—the much greater thickness of the uninjured gelatinous tissues serving to preserve the shape of the umbrella. But although the contractile sheet which lines the umbrella is thus completely severed throughout its whole diameter, it again reunites, or heals up, in from four to eight hours after the operation.

CHAPTER V.

EXPERIMENTS IN SECTION OF NAKED-EYED MEDUSÆ.

Distribution of Nerves in Sarsia.

My experiments have shown that the nervous system in the naked-eyed Medusæ is more highly organized, or integrated, than it is in the covered-eyed Medusæ; for whereas in the latter I obtained no evidence of the gathering together of nerve-fibres into definite bundles or trunks (the plexus being evenly distributed over the entire surface of the neuro-muscular sheet lining the umbrella), in the former I found abundant evidence of this advance in organization. And as the experiments in this connection serve to substantiate the histological researches of Professors Haeckel, Schultz, Eimer, and Hertwig, in as far as the distribution of the main nerve-trunks is concerned, I shall here detail at some length the character and results of these experiments in the case of Sarsia.

The occurrence of reflex action in Sarsia is of a very marked and unmistakable character. I may begin by stating that when any part of the internal surface of the bell is irritated, the manubrium responds; but as there is no evidence of ganglia occurring in the manubrium, this cannot properly be regarded as a case of reflex action. But now the converse of the above statement is likewise true, viz. that when any part of the manubrium is irritated, the bell responds; and it is in this that the unequivocal evidence of reflex action consists, for while the sympathy of the manubrium with the bell is not in the least impaired by removing the marginal ganglia of the latter, the sympathy of the bell with the manubrium is by this operation entirely destroyed.

We have thus very excellent demonstration of the occurrence of reflex action in the Medusæ. Further experiments show that the reflex action occurs, not between the marginal ganglia and every part of the manubrium, but only between the marginal ganglia and the point of the bell from which the manubrium is suspended—it being only the pull which is exerted upon this point when the manubrium contracts that acts as a stimulus to the marginal ganglia. But the high degree of sensitiveness shown by the marginal ganglia to the smallest amount of traction of this kind is quite as remarkable as their lack of sensitiveness to disturbances going on in the manubrium.

Turning now to the physiological evidence of the distribution of nerves in Sarsia, when one of the four tentacles is very gently irritated, it alone contracts. If the irritation be slightly stronger, all the four tentacles, and likewise the manubrium, contract. If one of the four tentacles be irritated still more strongly, the bell responds with one or more locomotor contractions. If in the latter case the stimulus be not too strong, or, better still, if the specimen operated on be in a non-vigorous or in a partly anæsthesiated state, it may be observed that a short interval elapses between the response of the tentacles and that of the bell. Lastly, the manubrium is much more sensitive to a stimulus applied to a tentacle, or to one of the marginal bodies, than it is to a stimulus applied at any other part of the nectocalyx.

These facts clearly point to the inference that nervous connections unite the tentacles with one another and also with the manubrium; or, perhaps more precisely, that each marginal body acts as a co-ordinating centre between nerves proceeding from it in four directions, viz. to the attached tentacle, to the margin on either side, and to the manubrium. This, it will be observed, is the distribution which Haeckel describes as occurring in Geryonia, and Schultz as occurring in Sarsia. It is, further, the distribution to which my explorations by stimulus would certainly point. But, in order to test the matter still more thoroughly, I tried the effects of section in destroying the physiological relations which I have just described. These effects, in the case of the tentacles, were sufficiently precise. A minute radial cut (only just long enough to sever the tissues of the extreme margin) introduced between each pair of adjacent marginal bodies completely destroyed the physiological connection between the tentacles. If only three marginal cuts were introduced, the sympathy between those two adjacent tentacles between which no cut was made continued unimpaired, while the sympathy between them and the other tentacles was destroyed.

The nervous connections between the tentacles and the manubrium are of a more general character than those described between the tentacles themselves; that is to say, severing the main radial nerve-trunks produces no appreciable effect upon the sympathy between the tentacles and the manubrium.

The nervous connections between the whole excitable surface of the nectocalyx and the manubrium are likewise of this general character, so that, whether or not the radial nerve-trunks are divided, the manubrium will respond to irritation applied anywhere over the internal surface of the nectocalyx. The manubrium, however, shows itself more sensitive to stimuli applied at some parts of this surface than it is to stimuli applied at other parts, although in different specimens there is no constancy as to the position occupied by these excitable tracts.

Distribution of Nerves in Tiaropsis Indicans.

We have seen that in Sarsia reflex action obtains between the manubrium and the nectocalyx; we shall now see that in Tiaropsis indicans something resembling reflex action obtains between the nectocalyx and the manubrium. The last-named species is a new one, which I have described elsewhere, and I have called it "indicans" from a highly interesting and important peculiarity of function which is manifested by its manubrium. The Medusa in question measures about one and a half inches in diameter, and is provided with a manubrium of unusual proportional size, its length being about five-eighths of an inch, and its thickness

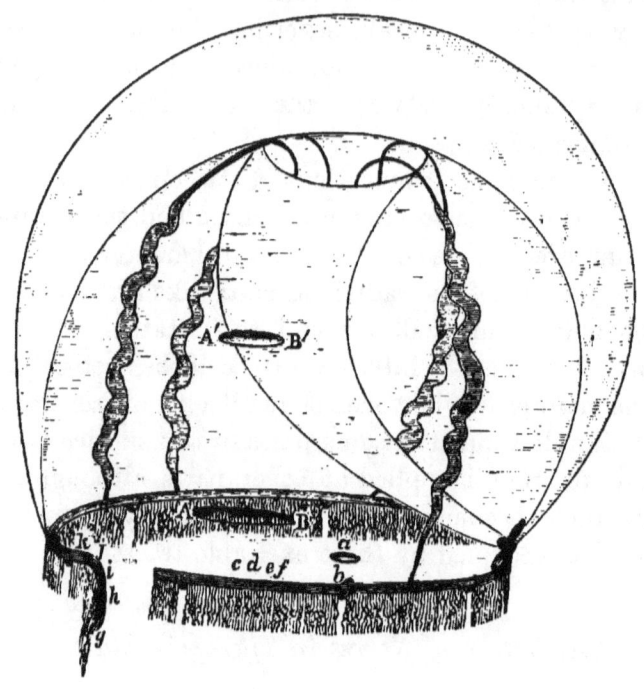

Fig. 22.

being also considerable. Now, if any part of the nectocalyx be irritated, the following series of phenomena takes place. Shortly after the application of the stimulus, the large manubrium suddenly contracts—the appearance presented being that of an exceedingly rapid crouching movement. The crouching attitude in which this movement terminates continues for one or two seconds, after which the organ begins gradually to resume its former dimensions. Concurrently with these movements on the part of the manubrium, the portion of the nectocalyx which has been stimulated bends inwards as far as it is able. The manubrium now begins to deflect itself towards the bent-in portion of the nectocalyx; and this deflection continuing with a somewhat rapid motion, the extremity of the manubrium is eventually brought, with unerring precision, to meet the in-bent portion of the nectocalyx. I here introduce a drawing of more than life-size to render a better idea of this *pointing* action by the manubrium to a seat of irritation located in the bell. It must further be stated that in the unmutilated animal such action is quite invariable, the tapered extremity of the manubrium never failing to be placed on the exact spot in the nectocalyx where the stimulation is being, or had previously been, applied. Moreover, if the experimenter irritates one point of the nectocalyx, with a needle or a fine pair of forceps for instance, and while the manubrium is applied to that point he irritates another point, then the manubrium will leave the first point and move over to the second. In this way the manubrium may be made to indicate successively any number of points of irritation; and it is interesting to observe that when, after such a series of irritations, the animal is left to itself, the manubrium will subsequently continue for a considerable time to visit first one and then another of the points which have been irritated. In such cases it usually dwells longest and most frequently on those points which have been irritated most severely.

I think the object of these movements is probably that of stinging the offending body by means of the urticating cells with which the extremity of the manubrium is armed. But, be the object what it may, the fact of these movements occurring is a highly important one in connection with our study of the distribution of nerves in Medusæ, and the first point to be made out with regard to these movements is clearly as to whether or not they are truly of a reflex character. Accordingly, I first tried cutting off the margin, and then irritating the muscular tissue of the bell; the movements in question were performed exactly as before. I was thus led to think it probable that the reflex centres of which I was in search might be seated in the manubrium. Accordingly, I cut off the manubrium, and tried stimulating its own substance directly. I found, however, that no matter how small a portion of this organ I used, and no matter from what part of the organ I cut it, this portion would do its best to bend over to the side which I irritated. Similarly, no matter how short a stump of the manubrium I left in connection with the nectocalyx, on irritating any part of the latter, the stump of the manubrium would deflect itself towards that part of the bell, although, of course, from its short length it was unable to reach it. Hence there can be no doubt that every portion of the manubrium—down, at least, to the size which is compatible with conducting these experiments—is independently endowed with the capacity of very precisely localizing a point of irritation which is seated either in its own substance or in that of the bell.

We have here, then, a curious fact, and one which it will be well to bear in mind during our subsequent endeavours to frame some sort of a conception regarding the nature of these primitive nervous tissues. The localizing function, which is so very efficiently performed by the manubrium of this Medusa, and which if anything resembling it occurred in the higher animals would certainly have definite ganglionic centres for its structural co-relative, is here shared equally by every part of the exceedingly tenuous contractile tissue that forms the outer surface of the organ. I am not aware that such a diffusion of ganglionic function has as yet been actually proved to occur in the animal kingdom, but I can scarcely doubt that future investigation will show such a state of things to be of common occurrence among the lower members of that kingdom.*

* The only case I know which rests on direct observation, and which is at all parallel to the one above described, is the case of the tentacles of Drosera. Mr. Darwin found, when he cut off the apical gland of one of these tentacles, together with a small portion of the apex, that the tentacle thus mutilated would no longer respond to stimuli applied directly to itself. Thus far the case differs from that of the manubrium of Tiaropsis indicans, and, in respect of localization of co-ordinating function, resembles that of ganglionic action. But Mr. Darwin also found that such a "headless tentacle" continued to be influenced by stimuli applied to the glands of neighbouring tentacles—the headless one in that

I shall now proceed to consider the nature of the nervous connections between the nectocalyx and manubrium of this Medusa.

Bearing in mind that in an unmutilated Tiaropsis indicans the manubrium invariably localizes with the utmost precision any minute point of irritation situated in the bell, the significance of the following facts is unmistakable, viz. that when a cut is introduced between the base of the manubrium and the point of irritation in the bell, the localizing power of the former, as regards that point in the latter, is wholly destroyed. For instance, if such a cut as that represented at a (see Fig. 22) be made in the nectocalyx of this Medusa, the manubrium will no longer be able to localize the seat of a stimulus applied below that cut, as, for instance, at b. Now, having tried this experiment a number of times, and having always obtained the same result, I conclude that the nervous connections between the nectocalyx and the manubrium, which render possible the localizing action of the latter, are connections the functions of which are intensely specialized, and the distribution of which is radial.

So far, then, we have highly satisfactory evidence of tissue-tracts performing the function of afferent nerves. But another point of interest here arises. Although, in the experiment just described, the

case bending over in whatever direction it was needful for it to bend, in order to approach the seat of stimulation. This shows that the analogue of ganglionic function must here be situated in at least more than one part of a tentacle; and I think it is not improbable that, if trials were expressly made, this function would be found to be diffused throughout the whole tentacle.

manubrium is no longer able to *localize* the seat of stimulation in the bell, it nevertheless continues able to perceive, so to speak, that stimulation is being applied in the bell *somewhere;* for every time any portion of tissue below the cut *a* is irritated, the manubrium actively dodges about from one part of the bell to another, applying its extremity now to this place and now to that one, as if seeking in vain for the offending body. If the stimulation is persistent, the manubrium will every now and then pause for a few seconds, as if trying to decide from which direction the stimulation is proceeding, and will then suddenly move over and apply its extremity, perhaps to the point that is opposite to the one which it is endeavouring to find. It will then suddenly leave this point and try another, and then another, and another, and so on, as long as the stimulation is continued. Moreover, it is important to observe that there are *gradations* between the ability of the manubrium to localize correctly and its inability to localize at all, these gradations being determined by the circumferential distance from the end of the cut and the point of stimulation. For instance, in Fig. 22, suppose a cut A B, quarter of an inch long, to be made pretty close to the margin and concentric with it, then a stimulus applied at the point *c*, just below the middle point of A B, would have the effect of making the manubrium move about to various parts of the bell, without being able in the least degree to localize the seat of irritation. But if the stimulus be applied at *d*, the manubrium will probably be so far able to localize the seat of irritation as to confine its movements, in its search for the offending body, to perhaps the *quadrant* of the bell in which the stimulation is being applied. If the stimulation be now supplied at e, the localization on the part of the manubrium will be still more accurate; and if applied at f (that is, *almost* beneath one end of the cut A B), the manubrium may succeed in localizing quite correctly.

These facts may also be well brought out by another mode of section, viz. by cutting round a greater or less extent of the marginal tissue, leaving one end of the resulting slip free, and the other end attached *in situ*. If this form of section be practised on Tiaropsis indicans, as represented at $g\ k$ in the figure, it may also be observed that irritation of a distant point in the marginal strip, such as g or h, causes the manubrium to move in various directions, without any special reference to that part of the bell which the irritated point of the marginal strip would occupy if *in situ*. But if the stimulation be applied only one or two millims. from the point of attachment of the marginal strip, as at i, the manubrium will confine its localizing motions to perhaps the proper quadrant of the bell; and if the stimulus be applied still nearer to the attachment of the severed strip, as at j, the localizing motions of the manubrium may become quite accurate.

Again, with regard to *radial* distance, if the cut A B in the figure were situated higher up in the bell, as at A' B', and the arc, c, d, e, f, of the margin irritated as before, the manubrium would be able to localize better than if, as before, the radial distance between A B and c, d, e, f were less. The greater this radial distance, the better would be the localizing power of the manubrium; so that, for instance, if the cut A′ B′ were situated nearly at the base of the manubrium, the latter organ might be able to localize correctly a stimulus applied, not only as before at f, but also at e or d. In such comparative experiments, however, it is to be understood that the higher up in the bell a cut is placed, the shorter it must be; for a fair comparison requires that the two ends of the cut shall always touch the same two radii of the nectocalyx. Still, if the cut is only a very short one (say one or two millims. long), this consideration need not practically be taken into account; for such a cut, if situated just above the margin, as represented at a, will have the effect of destroying the localizing power of the manubrium as regards the corresponding arc of the margin; but if situated high up in the bell, even though its length be still the same, it will not have this effect.

From all this, then, we have seen that the connections which render possible the *accurate* localizing functions of the manubrium are almost, though not quite, exclusively radial. We have also seen that between accurate localization and mere random movements on the part of the manubrium there are numerous gradations, the degree of decline from one to the other depending on the topographical relations between the point of stimulation and the end of the section (the section being of the form represented by A B in the figure). These relations, as we have seen, are the more favourable to correct localization: (*a*) the greater the radial distance between the point of stimulation and the end of the section; and (*b*) the less the circumferential distance between the point of the stimulation and the radius let fall from the end of the section. But we have seen that the limits as regards severity of section within which these gradations of localizing ability occur, are exceedingly restricted—a cut of only a few millims. in length, even though situated at the greatest radial distance possible, being sufficient to destroy all localizing power of the manubrium as regards the middle point of the corresponding arc of the margin, and a stimulus applied only a few millims. from the attached end of a severed marginal strip entirely failing to cause localizing action of the manubrium. Lastly, we have seen that even after all localizing action of the manubrium has been completely destroyed by section of the kinds described, this organ nevertheless continues actively, though ineffectually, to search for the seat of irritation.

The last-mentioned fact shows that after excitational continuity of a higher order has been destroyed, excitational continuity of a lower order nevertheless persists; or, to state the case in other words, the fact in question shows that after severance of the almost exclusively radial connections between the bell and the manubrium, by which the perfect or unimpaired localizing function of the latter is rendered possible, other connections between these organs remain which are not in any wise radial. I therefore next tested the degree in which these non-radial connections might be cut without causing destruction of that excitational continuity of a lower order which it is their function to maintain. It will here suffice to record one mode of section which has yielded definite results. A glance at the accompanying illustration (Fig. 23) will show the manner in which the Medusa is pre-

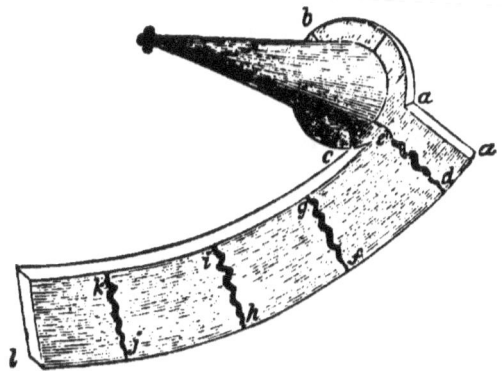

Fig 23.

pared. The margin having been removed (in order to prevent possible conduction by the marginal nerve-fibres), a single deep radial cut (*a a*) is first made, and then a circumferential cut (*a, b, c*) is carried nearly all the way round the base of the manubrium. In this way the nectocalyx, deprived of its margin, is converted into a continuous band of tissue, one of the ends of which supports the manubrium. Now it is obvious that this mode of section must be very trying to nervous connections of any kind subsisting between the bell and the manubrium. Nevertheless, in many cases, irritating any part of the band $a\,l$ has the effect of causing the manubrium to perform the active random motions previously described. In such cases, however, it is observable that the further away from the manubrium the stimulus is applied, the less active is the response of this organ. In very many instances, indeed, the manubrium altogether fails to respond to stimuli applied at more than a certain distance from itself. For example, referring to Fig. 23, the manubrium might actively respond to irritation of any point in the division d, e, f, g, while to irritation of any point in the division f, g, h, i its responses would be weaker, and to irritation of any point in h, i, j, k, they would be very uncertain or altogether absent. Hence in this form of section we have reached about the limit of tolerance of which the non-radial connections between the bell and manubrium are capable.

Another interesting fact brought out by this form of section is, that the radial tubes are tracts of comparatively high irritability as regards the manubrium; for the certainty and vigour with which the manubrium responds to a stimulus applied at one of the severed radial tubes, f, g, or h, i, or j, k, contrast strongly with the uncertainty and feebleness with which it often responds to stimuli applied between any of these tubes. Indeed, it frequently happens that a specimen which will not respond at all to a stimulus applied between two radial tubes, will respond at once to a stimulus applied much further from the manubrium, but *in the course of* the radial tube *f k*.

And this leads us to another point of interest. In such a form of section, when any part of the mutilated nectocalyx is irritated, the manubrium shows a very marked tendency to touch some point in the tissue-mass *a a d e* (Fig. 23) by which it still remains in connection with the bell, and through which, therefore, the stimulus must pass in order to reach the manubrium. And it is observable that this tendency is particularly well marked if the section has been planned as represented in Fig. 23, *i.e.* in such a way as to leave the tissue-tract *a a d e* pervaded by a nutrient-tube *d e*, this tube being thus left intact. When this is done, the manubrium most usually points to the uninjured nutrient-tube *d e* every time any part of the tissue-band *a l* is irritated.

Let us now very briefly consider the inferences to which these results would seem to point. The fact that the localizing power of the manubrium is completely destroyed as regards all parts of the bell lying beyond an incision in the latter, conclusively proves, as already stated, that all parts of the bell are pervaded by radial lines of differentiated tissue, which have at least for one of their functions the conveying of impressions to the manubrium. The fact in question also proves that the particular effect which is produced on the manubrium by stimulating any one of these lines cannot be so produced by stimulating any of the other lines. But although these tracts of differentiated tissue thus far resemble afferent nerves in their function, we soon see that in one important particular they differ widely from such nerves; for we have seen that, after they have been divided, stimulation of their peripheral parts still continues to be transmitted to their central parts, as shown by the non-localizing movements of the manubrium. Of course this transmission cannot take place through the divided tissue-tracts themselves; and hence the only hypothesis we can frame to account for the fact of its occurrence is that which would suppose these tissue-tracts, or afferent lines, to be capable of vicarious action. Such vicarious action would probably be effected by means of intercommunicating fibres, the directions of which would probably be various. In this way we arrive at the hypothesis of the whole contractile sheet being pervaded by an intimate plexus of functionally differentiated tissue, the constituent elements of which are capable of a vicarious action in a high degree.

Now we know from histological observation that there is a plexus of nerve-fibres pervading the whole expanse of the contractile sheet, and therefore we may conclude that this is the tissue through which the effects are produced. But, if so, we must further conclude that the fibres of this nerve-plexus are capable of vicarious action in the high degree which I have explained.

And this hypothesis, besides being recommended by the consideration that it is the only one available, is confirmed by the fact that the stimuli which it supposes to escape from a severed phalanx of nerve-fibres, and then to reach the manubrium after being diffused through many or all of the other radial lines (such stimuli thus converging from many directions), are responded to when they reach the manubrium, not by any decided localizing action on the part of the latter, but, as the hypothesis would lead us to expect, by the tentative and apparently random motions which are actually observed. Moreover, we must not neglect to notice that these tentative or random movements resemble in every way the localizing movements, save only in their want of precision. Again, this hypothesis is rendered more probable by the occurrence of those *gradations* in the localizing power of the manubrium which we have seen to be so well marked under certain conditions. The occurrence of such gradations under the conditions I have named is what the theory would lead us to expect, because the closer beneath a section that a stimulus is applied, the greater must be the immediate lateral spread of the stimulus through the plexus before it reaches the manubrium. Similarly, the further the circumferential distance from the nearest end of such a section that the stimulus is applied, the greater will be its lateral spread before reaching the manubrium. Lastly, the present hypothesis would further lead us to anticipate the fact that when Tiaropsis indicans is prepared as represented in Fig. 23, the manubrium refers a stimulus applied anywhere in the mutilated nectocalyx to the band of tissue by which it is still left in connection with that organ;

for it is evident that, according to the hypothesis, the radial fibres occupying such a band are the only ones whose irritation the manubrium is able to perceive, and hence it is to be expected that it should tend to refer to these particular fibres a source of irritation occurring anywhere in the mutilated bell.

It is not quite so easy to understand why, in the last-mentioned experiment, the manubrium should tend to refer a seat of irritation to the unsevered nutrient tube, or nerve-trunk, rather than to the unsevered nerves in the general nerve-plexus on either side of that nerve-trunk; for if this nerve-trunk at all resembles in its functions the nerve-trunks of higher animals, the afferent elements collected in it ought to communicate to the manubrium the impression of having had their *distal* terminations irritated, and therefore the fact of a number of such elements being collected into a single trunk ought not to cause the manubrium to refer a distant seat of irritation to that trunk rather than to any of the parts from which the plexus-elements may emanate. Concerning this difficulty, however, I may observe that we seem to have in it one of those cases in which it would be very unsafe to argue, with any confidence, from the highly integrated nervous systems with which we are best acquainted, to the primitive nervous systems with which we are now concerned. And although it would occupy too much space to enter into a discussion of this subject, I may further observe that I think it is not at all improbable that the manubrium of Tiaropsis indicans should, in the absence of more definite information, refer a distant seat of injury to that tract of collected afferent elements through which it actually receives the strongest stimulation.

Staurophora Laciniata.

This is a Medusa about the size of a small saucer which responds to stimulation of its marginal ganglia, or radial nerve-trunks, by a peculiar spasmodic movement. This consists in a sudden and violent contraction of the entire muscle-sheet, the effect of which is to draw together all the gelatinous walls of the nectocalyx in a far more powerful manner than occurs during ordinary swimming. In consequence of this spasmodic action being so strong, the nectocalyx undergoes a change in form of a very marked and distinctive character. The corners of the four radial tubes, being occupied by comparatively resisting tissue, are not so much affected by the spasm as are other parts of the bell; and they therefore constitute a sort of framework upon which the rest of the bell contracts, the whole bell thus assuming the form of an almost perfect square, with each side presenting a slight concavity inwards. These spasmodic movements, however, are quite unmistakable when they occur even in a very minute portion of detached tissue; for, however large or small the portion may be, when in a spasm it folds upon itself with the characteristically strong and persistent contraction. I say *persistent* contraction, because a spasmodic contraction, besides being of unusual strength, is also of unusual duration; that is to say, while an ordinary systolic movement only lasts a short time, a spasm lasts from six to ten seconds or more, and this whether it occurs in a large or in a small piece of tissue. Again, the diastolic movements differ very much in the case of an ordinary locomotor contraction and in that of a spasm; for while in the former case the process of relaxation is rapid even to suddenness, in the latter it is exceedingly prolonged and gradual, occupying some four or five seconds in its execution, and, from its slow but continuous nature, presenting a graceful appearance. Lastly, the difference between the two kinds of contraction is shown by the fact that, while a spasm is gradually passing off the ordinary rhythmical contractions may often be seen to be superimposed on it—both kinds of contraction being thus present in the same tissue at the same time.

Now the point with which we shall be especially concerned is, that it is only stimulation of *certain parts* of the organism which has the effect of throwing it into a spasm. These parts are the margin (including the tentacles) and the courses of the four radial tubes (including the manubrium, which in this species is spread over the radial tubes). This limitation, however, is not invariable; for I have often seen individuals of this species respond with a spasm to irritation of the general contractile tissue. Nevertheless, such response to such stimulation in the case of this species is exceptional—

the usual response to muscular irritation being an ordinary locomotor contraction, which forms a marked contrast to the tonic spasm that *invariably* ensues upon stimulation of the margin, and *almost* invariably upon the stimulation of a radial tube.

The first question I undertook to answer was the amount of section which the excitable tissues of Staurophora laciniata would endure without losing their power of conducting the spasmodic contraction from one of their parts to another. This was a very interesting question to settle, because Staurophora laciniata, like all the other species of discophorus naked-eyed Medusæ, differs from Aurelia, etc., in that the ordinary contraction-waves are very easily blocked by section. It therefore became interesting to ascertain whether or not the wave of spasm admitted of being blocked as easily. First, then, as regards the margin. If this be all cut off in a continuous strip, with the exception of one end left attached *in situ*, irritation of any part of the almost severed strip will cause a responsive spasm of the bell, so soon as the wave of stimulation has time to reach the latter. I next continued this form of section into the contractile tissues themselves, carrying the incision round and round the bell in the form of a spiral, as represented in the case of Aurelia by Fig. 11, page 70. In this way I converted the whole Medusa into a ribbon-shaped piece of tissue;* and on now stimulating the marginal

---

* It may be stated that while conducting this mode of section of Staurophora laciniata, the animal responds to each cut of the contractile tissues with a locomotor contraction (or it may not

tissue at one end of the ribbon, a portion of the latter would go into a spasm. The object of this experiment was to ascertain how far into the ribbon-shaped tissue the wave of spasm would penetrate. As I had expected, different specimens manifested considerable differences in this respect, but in all cases the degree of penetration was astonishingly great. For it was the exception to find cases in which the wave of spasm failed to penetrate from end to end of a spiral strip caused by a section that had been carried twice round the nectocalyx; and this is very astonishing when we remember that the ordinary contraction-waves, whether originated by stimulation of the contractile tissues or arising spontaneously from the point of attachment of the marginal strip, usually failed to penetrate further than a quarter of the way round. Moreover, these waves of spasm will continue to penetrate such a spiral strip even after the latter has been submitted to a system of interdigitating cuts of a very severe description.

Now, we have here to deal with a class of facts which physiologists will recognize as of a perfectly novel character. Why it should be that the very tenuous tracts of tissue which I have named should have the property of responding even to a feeble stimulus by issuing an impulse of a kind which throws the contractile tissues into a spasm; why it should be that a spasm, when so originated, should

respond at all); but each time the section crosses one of the radial tubes, the whole bell in front of the section, and the whole strip behind it, immediately go into a spasm.

manifest a power of penetration to which the normal contractions of the tissues in which it occurs bear so small a proportion; why it is that the contractile tissues should be so deficient in the power of originating a spasm, even in response to the strongest stimulation applied to themselves;—these and other questions at once suggest themselves as questions of interest. At present, however, I am wholly unable to answer them; though we may, I think, fairly assume that it is the ganglionic element in the margin, and probably also in the radial tubes, which responds to direct stimulation by discharging a peculiar impulse, which has the remarkable effect in question. For the sake of rendering the matter quite clear, let us employ a somewhat far-fetched but convenient metaphor. We may compare the general contractile tissues of this Medusa to a mass of gun-cotton, which responds to ignition (direct stimulation) by burning with a quiet flame, but to detonation (marginal stimulation) with an explosion. In the tissue, as in the cotton, every fibre appears to be endowed with the capacity of liberating energy in either of two very different ways; and whenever one part of the mass is made, by the appropriate stimulus, to liberate its energy in one of these two ways, all other parts of the mass do the same, and this no matter how far through the mass the liberating process may have to extend. Now, employing this metaphor, what we find is that, while the contractile fibres resemble the cotton fibres in the respects just mentioned, the ganglion cells resemble detonators, when themselves directly stimulated. In other words, the ganglion-cells of this Medusa are able to originate two very different kinds of impulse, according as they liberate their energy spontaneously or in answer to direct stimulation, and the muscular tissues respond with a totally different kind of contraction in the two cases. Possibly, indeed, direct stimulation of the ganglia is followed by a spasm of the muscular tissue only because a greater amount of ordinary ganglion influence is thus liberated than in the case of a merely spontaneous discharge. If this were the explanation, however, I should not expect so great a contrast as there is between the facility with which a spasm may be caused by stimulation of the margin and of the contractile tissue respectively. The slightest nip of the margin of Staurophora laciniata, for instance, is sufficient to cause a spasm, whereas even crushing the contractile tissues with a large pair of dissecting-forceps will probably fail to cause anything other than an ordinary contraction. Nevertheless, pricking the margin with a fine needle usually has the effect of causing only a locomotor contraction.

In conclusion, I may state that anæsthetics have the effect of blocking the spasmodic wave in any portion of tissue that is submitted to their influence. It is always observable, however, that this effect is not produced till after spontaneity has been fully suspended, and even muscular irritability destroyed as regards direct stimulation. Up to this stage the certainty and vigour of the spasm consequent on marginal irritation are not perceptibly impaired; but soon after this stage the intensity of the spasm begins to become less, and later still it assumes a *local* character. It is important, also, to notice that at this stage the effect of marginal stimulation is very often that of producing a *general locomotor* contraction, and sometimes a series of two or three such. During recovery in normal sea-water all these phases occur in reverse order.

CHAPTER VI.

CO-ORDINATION.

Covered-eyed Medusæ.

FROM the fact that in the covered-eyed Medusæ the passage of a stimulus-wave is not more rapid than that of a contraction-wave, we may be prepared to expect that in these animals the action of the locomotor ganglia is not, in any proper sense of the term, a co-ordinated action; for if a stimulus-wave cannot outrun a contraction-wave, one ganglion cannot know that another ganglion has discharged its influence till the contraction-wave, which results from a discharge of the active ganglion, has reached the passive one. And this I find to be generally the case; for it may usually be observed that one or more of the lithocysts are either temporarily or permanently prepotent over the others, *i.e.* that contraction-waves emanate from the prepotent lithocysts, and then spread rapidly over the swimming-bell, without there being any signs of co-ordinated or simultaneous action on the part of the other lithocysts. Nevertheless, in many cases such prepotency cannot, even with the greatest care, be observed; but upon every pulsation all parts of the swimming-bell seem to contract at the same instant. And this apparently perfect co-ordination among the eight marginal ganglia may continue for any length of time. I believe, however, that such apparently complete physiological harmony is not co-ordination properly so called, *i.e.* is not due to special nervous connections between the ganglia; for, if such were the case, perfectly synchronous action of this kind ought to be the rule rather than the exception.

I am therefore inclined to account for these cases of perfectly synchronous action by supposing that all, or most, of the ganglia require exactly the same time for their nutrition; that they are, further, of exactly equal potency in relation to the resistance (or excitability) of the surrounding contractile tissues; and that, therefore, the balance of forces being exactly equal in the case of all, or most, of the ganglia, their rhythm, though perfectly identical, is really independent. I confess, however, that I am by no means certain regarding the accuracy of this conclusion, as it is founded on negative rather than on positive considerations: that is to say, I arrive at this conclusion regarding the cases in which such apparent co-ordination is observable only because in other cases such apparent co-ordination is not observable; and also, I may add, because my experiments in section have not revealed any evidence of nervous connections capable of conducting a stimulus-wave with greater rapidity than a contraction-wave. I therefore consider this conclusion an uncertain one, and its uncertainty is, perhaps, still further increased by the result of the following experiments.

If a covered-eyed Medusa be chosen in which perfectly synchronous action of the ganglia is observable, and if a deep radial incision be made between each pair of adjacent ganglia—the incisions being thus eight in number and carried either from the margin towards the centre or *vice versâ*—it then becomes conspicuous enough that the eight partially divided segments no longer present synchronous action; for now one segment and now another takes the initiative in starting a contraction-wave, which is then propagated to the other segments. And it is evident that this fact tends to negative the above explanation, for if the discharges of the ganglia are independently simultaneous before section, we might expect them to continue so after section. It must be remembered, however, that the form of section we are considering is a severe one, and that it must therefore not only give rise to general shock, but also greatly interfere with the passage of contraction-waves, and, in general, disturb the delicate conditions on which, according to the suggested explanation, the previous harmony depended. Besides, as we shall subsequently see, for some reason or other segmentation of a Medusa profoundly modifies the rate of its rhythm. In view of these considerations, therefore, the results yielded by such experiments must not be regarded as having any conclusive bearing on the question before us; and as these or similar objections apply to various other modes of section by which

I have endeavoured to settle this question, I will not here occupy space in detailing them.

It seems desirable, however, in this connection again to mention a fact briefly stated in a former chapter, namely, that section conclusively proves a contraction-wave to have the power, when it reaches a lithocyst, of stimulating the latter into activity; for it is not difficult to obtain a series of lithocysts connected in such a manner that the resistance offered to the passage of the waves by a certain width of the junction-tissue, is such as just to allow the residuum of the contraction-wave which emanates from one lithocyst to reach the adjacent lithocyst, thus causing it to originate another wave, which, in turn, is just able to pass to the next lithocyst in the series, and so on, each lithocyst in turn acting like a reinforcing battery to the passage of the contraction-wave. Now this fact, I think, sufficiently explains the mechanism of ganglionic action in those cases where one or more lithocysts are prepotent over the others; that is to say, the prepotent lithocyst first originates a contraction-wave, which is then successively reinforced by all the other lithocysts during its passage round the swimming-bell. In this way the passage of a contraction-wave is no doubt somewhat accelerated; for I found, in marginal strips, that the rate of transit from a terminal lithocyst to the other end of the strip was somewhat lowered by excising the seven intermediate lithocysts.

I may here state, in passing, a point of some little interest in connection with this reinforcing action of lithocysts. When I first observed this action, it appeared to me a mysterious thing why its result was always to propagate the contraction-wave in only one direction—the direction, namely, in which the wave happened to be passing before it reached the lithocyst. For instance, suppose we have a strip A D, with a lithocyst at each of the equidistant points A, B, C, D; and suppose that the lithocyst B originates a stimulus: the resulting contraction-wave passes, of course, with equal rapidity in the two opposite directions, B A, B C (arrows $b\,a, b\,c$): the contraction-wave $b\,a$ therefore

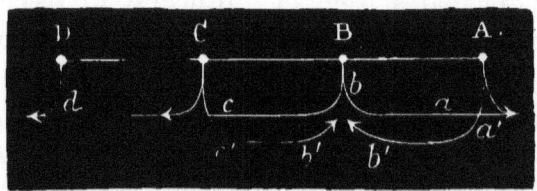

Fig. 24.

reaches the lithocyst A at the same time as the contraction-wave $b\,c$ reaches the lithocyst C, and so both A and C discharge simultaneously. What, then, should we expect to be the result? I think we should expect the wave $b\,c$ to continue on its course to D, after having been strengthened at C, and a *reflex* wave $a'\,b'$ to start from A (owing to the discharge at A), which would reach B at the same time as a similar *reflex* wave $c'\,b'$ starting from C (owing to the discharge at C); so that by the time the original wave $b\,c\,d$ had reached D, the point B would be the seat of a collision between the two reflex waves $a'\,b'$ and $c'\,d'$. And, not to push the supposed case further, it is evident that if such reflex waves were to occur, the resulting confusion would very soon require to end in tetanus. As a matter of fact, these reflex waves do not occur; and the question is, why do they not? Why is it that a wave is only reinforced in the direction in which it happens to be travelling—so that if, for instance, it happens to start from A in the above series, it is successively propagated by B C in the direction A, B, C, D, and in that direction only; whereas, if it happens to start from D, it is propagated by the same lithocysts in the opposite direction, D, C, B, A, and in that direction only—the wave in the one case terminating at the lithocyst D, and in the other case at the lithocyst A? Now, although this absence of reflex waves appears at first sight mysterious, it admits of an exceedingly simple explanation. I find that the contractile tissues of the covered-eyed Medusæ cannot be made to respond to two successive stimuli of minimal, or but slightly more than minimal intensity, unless such stimuli are separated from one another by a certain considerable interval of time. Now, when in the above illustration the contraction-wave starts from A, by the time it reaches B the portion of tissue included between A and B has just been in contraction in response to the stimulus from A, while the portion of tissue included between B and C has not been in contraction. Consequently, the stimulus resulting from a ganglionic discharge being presumably of minimal, or but slightly more than minimal intensity, the tissue included between A and B will not respond to the discharge of B; while the tissue included between B and C, not having been just previously in contraction, will respond. And conversely, of course, if the contraction-wave had been travelling in the opposite direction.

Seeing that this explanation is the only one possible, and that it moreover follows as a deductive necessity from my experiments on stimulation, I think there is no need to detail any of the further experiments which I made with the view of confirming it. But the following experiment, devised to confirm this explanation, is of interest in itself, and on this account I shall state it. Having prepared a contractile strip with a single remaining lithocyst at one end, I noted the rhythm exhibited by this lithocyst, and then imitated that rhythm by means of single induced shocks thrown in with a key at the other end of the strip. The effect of these shocks was, of course, to cause the contraction-waves to pass in the direction opposite to that in which they passed when originated by the lithocyst. Now I found, as I had expected, that so long as I continued exactly to imitate the rate of ganglionic rhythm, so long did the waves always pass in the direction B A—A being the lithocyst, and B the other end of the strip. I also found that if I allowed the rate of the artificially caused rhythm to sink slightly below that of the natural rhythm, after every one to six waves (the number depending on the degree in which the rate of succession of my induction shocks approximated to the rate of the natural rhythm) which passed from B to A, one would pass from A to B.*

Of course the only interpretation to be put on these facts is that every time an artificially started wave reached the terminal ganglion it caused the latter to discharge; but that the occurrence of a discharge could not in this case be rendered apparent, because of the inadequacy of that discharge to start a reflex wave. But that such discharges always took place was manifest, both *à priori* because from analogy we may be sure that if there had happened to be any contractile tissue of appropriate width on the other side of the ganglion, the discharge of the latter would have been rendered apparent, and *à posteriori* because, after the arrival of every artificially started wave, the time required for the ganglion to originate another wave was precisely the same as if it had itself originated the previous wave.

In view of these results, it occurred to me as an interesting experiment to try the effect on the natural rhythm of exhausting a ganglion thus situated, by throwing in a great number of shocks at the other end of the strip. I found that after five hundred single shocks had been thrown in with a rapidity almost sufficient to tetanize the strip, immediately after the stimulation ceased, the natural

---

* When two such waves met, they neutralized each other at their line of collision; or perhaps more correctly, the tissue on each side of that line, having just been in contraction, was not able again to convey a contraction-wave passing in the opposite direction to the wave which it had conveyed immediately before.

rhythm of the ganglion, which had previously been twenty in the minute, fell to fourteen for the first minute, eighteen for the second, and the original rate of twenty for the third. In such experiments the diminution of rate is most conspicuous during the first fifteen or thirty seconds of the first minute. Sometimes there are no contractions at all for the first fifteen seconds after cessation of the stimulating process, and in such cases the natural rhythm, when it first begins, may be as slow as one-half or even one-quarter its normal rate. All these effects admit of being produced equally well, and with less trouble, by faradizing the strip, when it may be even better observed how prolonged may be the stimulation, without causing anything further than such slight exhaustion of the ganglion as the above results imply.*

Naked-eyed Medusæ.

It would be impossible to imagine movements on the part of so simple an organism more indicative of physiological harmony than are the movements of Sarsia. One may watch several hundreds of these animals while they are swimming about in the same bell-jar and never perceive, as in the covered-eyed Medusæ, the slightest want of gang-

* In this description I have everywhere adopted the current phraseology with regard to ganglionic action—a phraseology which embodies the theory of ganglia supplying interrupted stimulation. But although I have done this for the sake of clearness, of course it will be seen that the facts harmonize equally well with the theory of continuous stimulation, to which I shall allude further on.

lionic co-ordination exhibited by any of the specimens. Moreover, that the ganglionic co-ordination is in this case wonderfully far advanced is proved by the fact of members of this genus being able to steer themselves while following a light, as previously described.*

In the discophorous species of naked-eyed Medusæ, however, perfectly co-ordinated action is by no means of such invariable occurrence as it is in Sarsia; for although in perfectly healthy and vigorous specimens systole and diastole occur at the same instant over the whole nectocalyx, this harmoniously acting mechanism is very liable to be thrown out of gear, so that when the animals are suffering in the least degree from any injurious conditions, often too slight and obscure to admit of discernment, the swimming movements are no longer synchronous over the whole nectocalyx; but now one part is in systole while another part is in diastole, and now several parts may be in diastole while other parts are in systole. And as in these animals very slight causes seem sufficient thus to impair the ganglionic co-ordination, it generally happens that in a bell-jar containing a number of specimens belonging to different species, numerous examples of more or less irregular swimming movements are observable.

Taking, then, the case of Sarsia first, from my

* Removing the manubrium does not interfere with this steering action; but if any considerable portion of the margin is excised, the animal seems no longer able to find the beam of light, even though one or more of the marginal bodies be left *in situ*.

previous observations on the physiological harmony subsisting between the tentacles, I was led to expect that the co-ordination of the locomotor ganglia was probably effected by means of the same tissue-tracts through which the intertentacular harmony was effected, namely, those situated in the margin of the bell. Accordingly, I introduced four short radial cuts, one midway between each pair of adjacent marginal bodies. The co-ordination, however, was not perceptibly impaired. I therefore continued the radial cuts, and found that when these reached one-half or two-thirds of the way up the sides of the inner bell (or contractile sheet), the co-ordination became visibly affected, and this for the first time.

I also tried the following experiment. Instead of beginning the radial cuts from the margin, I began them from the apex of the cone; and I found that however many of such cuts I introduced, and however far down the cone I carried them, so long as I did not actually sever the margin, so long did all the divisions of the bell continue to contract simultaneously.* This fact, therefore, proves that the margin of the bell is alone sufficient to maintain co-ordination.

* This could be particularly well seen if, after the extreme apex of the cone had been removed, one of the four radial cuts was continued through the margin, and the latter was then spread out into a linear form by gently pressing the animal against the flat side of the glass vessel in which it was contained. The same experiment performed on Aurelia is, of course, attended with a totally different result, now one segment and now another originating a discharge which then spreads to all the others in the form of a contraction-wave.

The next experiment I tried was to make four short radial incisions in the margin as before described, and then to continue *one* of these incisions the whole way up the bell. By careful observation I could now perceive that all the marginal ganglia did not discharge simultaneously; for when those situated nearest to the long radial cut happened to take the initiative, the resulting contraction-wave, having double the distance to travel which it would have had if the long radial cut had been absent, could now be followed by the eye in its very rapid course round the bell. Now, the fact that in this form of section I was able to detect the passage of a *wave*, proves that the three short radial sections had destroyed the co-ordinated action of the marginal ganglia.

From these experiments, then, I conclude that in this genus ganglionic co-ordination, in the strict sense of the term, is effected exclusively by means of the marginal nerves. And as these experiments on Sarsia are exceedingly difficult to conduct, owing to the very rapid passage of contraction-waves in this genus, it is satisfactory to find that this conclusion is further supported by the analogy which the other species of naked-eyed Medusæ afford, and to the consideration of which we shall now proceed.

The effects of four short radial incisions through the margin of any species of Tiaropsis, Thaumantias, Staurophora, etc., are usually very conspicuous. Each of the quadrants included between two adjacent incisions shows a strong tendency to assume an independent action of its own. This tendency is sometimes so pronounced as to amount almost to a total destruction of contractional continuity between two or more quadrants of the bell; but more usually the effect of the marginal sections is merely that of destroying excitational continuity, or at least physiological harmony.

It is an interesting thing that this form of section, although in actual amount so very slight, is attended with a much more pernicious influence on the vitality of the organism than is any amount of section of the general contractile tissues. Thus, if a specimen of Tiaropsis, for example, be chosen which is swimming about with the utmost vigour, and if four equidistant radial cuts only just long enough to sever the marginal canal be made, the animal will soon begin to show symptoms of enfeeblement, and within an hour or two after the operation will probably have ceased its swimming motions altogether. The animal, however, is not actually dead; for if while lying motionless at the bottom of the vessel it be gently stimulated, it will respond with a spasm as usual, and perhaps immediately afterwards give a short and feeble bout of swimming movements. These surprisingly pernicious results are not so conspicuous in the case of Sarsia, although in this genus likewise they are sufficiently well marked to be unmistakable. I here append a table to show the comparative effects of the operation in question on different species. The cases may be regarded as very usual ones, though it often happens that a longer time after the operation must elapse before the enfeebling effects become so pronounced.

Name of species.	Number of contractions during five minutes be.ore operation.	Number during one minute aft r operation.	Number during five minutes after operation.	Ultimate effects.
Tiaropsis diademata	57	11	0	Permanent rest.
—— indicans ...	148	23	0	,,
—— polydiademata	102	18	0	,,
—— oligoplocama	131	39	0	,,
Sarsia tubulosa ...	144	56	14	,,

This decided effect of so slight a mutilation will not, perhaps, appear to other physiologists so noteworthy as it appears to me; for no one who has not witnessed the experiments can form an adequate idea of the amount of mutilation of any parts, other than their margins, which the Medusæ will endure without even suffering from the effects of shock. Another point worth mentioning with regard to the operation we are considering is, that not unfrequently the interruptions of the margin, which have been produced artificially, begin to extend themselves through the nectocalyx in a radial direction; so that in some cases this organ becomes spontaneously segmented into four quadrants, which remain connected only by the apical tissue of the bell. I do not think that this is due to the mere mechanical tearing of the tissues as a consequence of the swimming motions, for the latter seem too feeble to admit of their producing such an effect.

In conclusion, I may state that I have been able temporarily to destroy the ganglionic co-ordination of Sarsiæ, by submitting the animals to severe nervous shock. The method I employed to produce the nervous shock, without causing mutilation, was to take the animal out of the water for a few seconds while I laid it on a small anvil, which I then struck violently with a hammer. On immediately afterwards restoring the Medusa to sea-water, spontaneity was found to have ceased, while irritability remained. After a time spontaneity began to return, and its first stages were marked by a complete want of co-ordination; soon, however, co-ordination was again restored. But this experiment by no means invariably yielded the same result. Spontaneity, indeed, was invariably suspended for a time; but its first return was not invariably, or even generally, marked by an absence of co-ordination, even though I had previously struck the anvil a number of times in succession. I was therefore led to try another method of producing nervous shock, and this I found a more effectual method than the one just described. It consisted in violently shaking the Sarsiæ in a bottle half filled with sea-water. I was surprised to find how violent and prolonged such shaking might be without any part of the apparently friable organism, except perhaps the tentacles and manubrium, being broken or torn. The subsequent effects of shock were remarkable. For some little time after their restoration to the bell-jar, the Sarsiæ had lost, not only their spontaneity, but also their irritability, for they would not respond even to the strongest stimulation. In the course of a few minutes, however, peripheral irritability returned, as shown by responses to nipping of the neuro-muscular sheet. The animals were now in the same condition as when anæsthesiated by caffein or other central nerve-poison; but in a few minutes later central or reflex irritability also returned, as shown by single responses to single nippings of the tentacles. Last of all spontaneity began to return, and was in some few cases conspicuously marked by a want of co-ordination, all parts of the margin originating impulses at different times, with the result of producing a continuous flurried or shivering movement of the nectocalyx. After a time, however, these movements became co-ordinated; but in most cases when a swimming bout had ended and a pause intervened, the next swimming bout was also inaugurated by a period of shivering before co-ordination became established. This effect might last for a long time, but eventually it, too, disappeared, the swimming bouts then beginning with co-ordinated action in the usual way.

CHAPTER VII.

NATURAL RHYTHM.

It will be convenient here to introduce all the observations that I have been able to make with regard to the natural rhythm of the Medusæ. As Dr. Eimer has also made some observations in this connection, before proceeding with the fresh points having relation to this subject, I shall consider those to which he alludes.

In Aurelia aurita, as Dr. Eimer noticed, the rate of the rhythm has a tendency to bear an inverse proportion to the size of the individual. Size, however, is far from being the only factor in determining the differences between the rate of the rhythm of different specimens, the individual variations in this respect being very great even among specimens of the same size. What the other factors in question may be, however, I am unable to suggest.

Dr. Eimer also affirms that the duration of the natural pauses, which in Aurelia habitually alternate with bouts of swimming, bears a direct proportion to the number and strength of the contractions that occurred in the previous bout of swimming. I observed that Sarsiæ are much better adapted than Aureliæ for determining whether any such precise relation obtains; for, in the first place, the strength of the contraction is more uniform, and, in the next place, the alternation of pauses with bouts of swimming is of a more decided character in Sarsiæ than in healthy specimens of Aureliæ. I further observed that in Sarsia no such precise relation did obtain, although in a very general way it is true, as might be expected, that unusually prolonged bouts of swimming were sometimes followed by pauses of unusual duration. As all the observations are very much the same, I shall only quote two of them:—

Sarsia.		*Sarsia* (another specimen).	
Number of pulsations.	Seconds of rest.	Number of pulsations.	Seconds of rest.
54	90	40	60
20	15	29	90
9	92	32	132
51	40	33	92
38	60	18	59
1	43	8	63
63	45	15	35
1	14	2	85
60	15	11	63
6	50	30	33
38	50	17	81
22	32	19	67
25	12	3	65
56	55	19	36
65	20	41	123
42	15	80	23
35	40	61	150
76	43	45	115
		40	120
		10	97
		14	35

These observations may be taken as samples of others which it would be unnecessary to quote, as it will be seen from the above that there is no precise relation between the number of the pulsations and the duration of the pauses. Nevertheless, that there is a general relation may be seen from some cases in which unusually prolonged pauses occur. The following instance will serve to show this:—

Sarsia (another specimen).

Number of pulsations.	Seconds of rest.
38	30
22	35
49	40
30	45
46	20
2	15
24	380
112	20
45	185
894	30
6	45
4	140
2	185
30	210
200	60

In this case, the relation between the long pause of 380 seconds and the subsequent prolonged swimming bout of 112 pulsations is obvious; also, as the latter was then followed by a short pause of twenty seconds and another comparatively short bout of forty-five pulsations, the refreshing influence of the previous 380 seconds rest may be supposed to have been not quite neutralized by the exhausting effect of the foregoing 112 pulsations. At any rate, looking to the general nature of the previous proportions (viz. in their sum $\frac{184}{211}$), it is certain that $\frac{380}{112}$ leaves a large preponderance in favour of nutrition, which preponderance is not much modified by adding the next succeeding proportion, thus, $\frac{380+20}{112+45} = \frac{400}{157}$. Consequently, the organism may fairly be supposed to have entered upon the next prolonged period of rest (viz. 185 seconds) with a large balance of reserve power; so that when to this large balance there was added the further accumulation due to the further rest of 185 seconds, we are not surprised to find the next succeeding swimming bout comprising the enormous number of 894 pulsations. But this great expenditure of energy seems to have been somewhat in excess of the energy previously accumulated by the prolonged rest, for this unusual expenditure seems next to have entailed an unusually prolonged period of exhaustion. At any rate, it is plainly observable that the next succeeding proportions are greatly in favour of repose; for it is not until 360 seconds have elapsed, with only twelve pulsations in the interval, that energy enough has been accumulated to cause a moderate bout of thirty pulsations. But next another long and sustained pause of 240 seconds supervenes, and, the animal being now fully refreshed with a large surplus of accumulated energy, the next succeeding swimming bout comprises two hundred pulsations. Lastly, there succeeded sixty seconds of rest, and here the observation terminated.*

* If the reader takes the trouble to ascertain the average proportion between the number of pulsations and the seconds of rest in the first observations as far down as the first long pause, viz. as above stated, $\frac{400}{157}$, and if he then balances the succeeding income

Effects of Segmentation on the Rhythm.

We have next to consider Dr. Eimer's observations concerning the effects on the rhythm of Aurelia which result on cutting the animal into segments; and here, again, I much regret to say that I cannot wholly agree with this author. He says he found evidence of a very remarkable fact, viz. that by first counting the natural rhythm of an unmutilated Aurelia, and then dividing the animal into two halves, one of these halves into two quarters, and one of these quarters into two eighths; the sum of the contractions performed by these four segments in a given time was equal to the number which had previously been performed in a similar time by the unmutilated animal. And not only so, but the number of contractions which each segment contributed to this sum was a number that stood in direct proportion to the size of the segment; so that the half contracted half as many times, the quarter a quarter as many times, and the eighth parts one-eighth part the number of times that the unmutilated Aurelia had previously contracted in a period of equal duration. I am glad to observe that Dr. Eimer does not regard this rule otherwise

and expenditure of energy of all the rest of the observations, he will find the net result to accord very precisely with the proportion he previously obtained. But, as already stated, any such precision as this is certainly the exception rather than the rule.

It may here be stated that after the sixty seconds of rest above recorded, the animal began another swimming bout. It was then immediately bisected, and the subsequent observations are detailed in the next footnote.

than as liable to frequent exception; for, as already observed, I cannot say that my experiments have tended to confirm it. I am only able to say that there is general tendency for the smaller segments of an Aurelia divided in this way to contract less frequently than the larger segments.

It would be tedious and unnecessary to quote any observations in this connection; but as these observations brought out very clearly a fact which I had previously suspected, I may detail one experiment to illustrate this point. The fact in question is, that the *potency of the lithocysts* in any given segment of a divided Aurelia has more to do with the frequency of its pulsations than has the size of the segment. As previously mentioned, one or more lithocysts may often be observed to be permanently prepotent over the others; and I may here observe that the segmentation experiments just described have shown the converse to be true, viz. that one or more lithocysts are often permanently feebler than the others. Well, if a specimen of Aurelia exhibiting decided prepotency in one or more of its lithocysts be watched for a considerable length of time, so as to be sure that the prepotency is not of a merely temporary character, and if the animal be then divided into segments in such a way that the prepotent lithocysts shall occupy the smaller segments, it may be observed, provided time be left for the tissues to recover, that the segments containing the prepotent lithocysts, notwithstanding their smaller size, contract more frequently than do the larger segments. Conversely, if the larger segments happen to contain feeble lithocysts, their contractions will be but few. I have, indeed, seen cases in which the lithocysts appeared to be quite functionless, so far as the origination of stimuli was concerned.

The following observations were made on a healthy specimen of Aurelia having all its lithocysts in good condition, but prepotency being well marked in the case of one of them, and also, though in a lesser degree, in the case of another. I divided the animal so as to leave one of these two prepotent lithocysts in each of the eighth-part segments, and the next most powerful lithocysts in the quadrant segment. In the following description, I shall call the two eighth-part segments A and B, the former letter designating the segment containing the most powerful lithocyst. The Aurelia before being divided manifested for several hours a very regular and sustained rhythm of thirty-two per minute. After its division, the various segments contracted at the following rates, in one-minute intervals:—

Time after operation.	Segment $\frac{1}{2}$.	Segment $\frac{1}{4}$.	Segment $\frac{1}{8}$ A.	Segment $\frac{1}{8}$ B.
½ hour.	20	25	27	15
1 ,,	20	25	27	15
2 hours.	29	25	27	16
4 ,,	19	16	27	12

Next morning, the water which contained the segments was somewhat foul, and this, as is always the case, gave rise to abnormally long pauses. This effect was much more marked in the case of some of the segments than in that of others. I therefore observed the segments over five-minute intervals, instead of one-minute intervals as on the previous day. The following is a sample of several observations, all yielding the same general result.

Segment $\frac{1}{2}$.		Segment $\frac{1}{4}$.	Segment $\frac{1}{8}$ A.	Segment $\frac{1}{8}$ B.
Number of pulsations.	Seconds of rest.	No motion during the hour of observation.	Continued persistently to contract with a nearly perfect rhythm of 78 in the 5 minutes during the hour of observation.	Rhythm tolerably perfect at 78 in the 5 minutes; but this was occasionally interrupted by long pauses of 4 or 5 minutes' duration.
12	120			
3	10			
2	20			
44	130			
12	20			
73	5 min.			
Average rate $14\frac{3}{5}$ per minute.		No motion.	Continuous rhythm at the rate of $15\frac{3}{5}$ per minute.	Interrupted rhythm at the rate of $15\frac{3}{5}$ per minute.

I now transferred all the segments to fresh sea-water, with the following results:—

Rhythm during first quarter of an hour immediately after transference, in five-minute intervals.

Time.	Segment $\frac{1}{2}$.	Segment $\frac{1}{4}$.	Segment $\frac{1}{8}$ A.	Segment $\frac{1}{8}$ B.
First 5 minutes.	139 (irregular).	0	83 (regular).	20 (irregular).
Second 5 minutes.	0	0	68 "	75 (regular).
Third 5 minutes.	100 (regular).	39 (irregular).	70 "	69 "

Rhythm two hours after transference (five-minute intervals).

Segment $\frac{1}{2}$.	Segment $\frac{1}{4}$.	Segment $\frac{1}{8}$ A.	Segment $\frac{1}{8}$ B.
82 (regular).	77 (regular).	70 (regular).	62 (regular).

Rhythm next day (five-minute intervals).

Segment ½.	Segment ¼.	Segment ⅛ A.	Segment ⅛ B.
68	55	17	Dead.

Next day all the segments were dead except the largest one, in which a single lithocyst still continued to discharge at the rate of twenty-four in five minutes.

Now, with regard to these tables, it is to be observed that during the first day the prepotent lithocyst in the eighth-part segment A maintained an undoubted supremacy over all the others, and that the same is true of the comparatively potent lithocysts in the quadrant. (This is not the case with segment B; probably the degree of prepotency of the lithocyst in this case was not sufficient to counteract the antagonistic influence of the small size of the segment.) But next day the supremacy of the small segment A was not so marked; for although its rhythm was more regular in the stale water than was that of the largest segment, its actual number of contractions in a given time was just about equal to that of the largest segment. Again, after transference to fresh sea-water, the balance began to fall on the side of the larger segments; for even the quadrant, which in the stale water had ceased its motions altogether, now held a middle position between that of the half-segment and the prepotent eighth-part segment. On the next day, again, the balance fell decidedly in favour of the larger segments, and the weaker eighth-part segment died. Lastly, next day all the smaller segments were dead.

Hence the principal facts to be gathered from these observations are, that as time goes on the rhythm of all the segments progressively decreases, and that the decrease is more marked in the case of the smaller than in that of the larger segments. This lesser endurance of the smaller segments also finds its expression in their earlier death. Now as these smaller segments started with a greater proportional amount of ganglionic power than the larger segments, their lesser amount of endurance can only, I think, be explained by supposing that the process of starvation proceeds at a rate inversely proportional to the size of the segment, a supposition which is rendered probable if we reflect that the smaller the segment the greater is the proportional area of severed nutrient tubes.* And

* It may be thought that the greater area of general tissue-mass in the larger segments than in the smaller, and not the lesser proportional area of tube-section, is the cause of the larger segments living longer than the smaller ones. I am led, however, to reject this hypothesis, because in Sarsia, where segmentation entails a comparatively small amount of tube-section, there is no constant rule as to the larger segments showing more endurance than the smaller ones—the converse case, in fact, being of nearly as frequent occurrence. I can only account for this fact by supposing that the endurance of the segments of Sarsia is determined by the degree in which the three or four minute open tube-ends become accidentally blocked. This supposition is the only one I can think of to account for the astonishing contrasts as to endurance that are presented by different segments of the same individual, and, I may add, of different individuals when deprived of their margins and afterwards submitted to the same conditions. For instance, a number of equally vigorous specimens had their margins removed, and were then suspended in a glass cage

in this connection it is interesting to observe that, although the endurance of the smaller segments attached to a buoy in the sea. Four days afterwards some of the specimens were putrid, while others were as fresh as they were when first operated on. Again, as an instance of the experiments in segmentation of Sarsia, I may quote an experiment in which a score of specimens were divided in all sorts of ways, such as leaving the manubrium attached to one half, or three marginal bodies in one portion and the remaining marginal body in the other portion, etc. Yet, although it was very exceptional to find the two portions presenting an equal degree of endurance, no uniform results pointing to the cause of the variations could be obtained. In most cases, however, the energy, as distinguished from the endurance of the larger segments, was conspicuously greater than that of the smaller. (But it is curious that in many cases the effects of *shock* appeared to be more marked in the larger than in the smaller segments—the latter, for some time after the operation, contracting much more frequently than the former.) To show both these effects, one experiment may be quoted. A specimen of Sarsia was divided into two parts, of which one was a quadrant.

Immediately after the operation the results were as follows:—

Portion $\frac{1}{4}$.		Portion $\frac{3}{4}$.	
Number of pulsations.	Minutes of rest.	Number of pulsations.	Minutes of rest.
20	0	0	5
4	4	10	2
15	5	46	1
6	3	23	2
		49	1
		900	1
45	12	117	1
		1145	13

To show the difference between the *endurance* of two halves of a bisected specimen of Sarsia, I may quote one experiment which

was less than that of the larger as regards the deprivation of nutriment, it was greater than that of

was performed on the same specimen as the one mentioned in the text to show the general relationship between the duration of the pauses and that of swimming bouts. (See last foot-note.)

Immediately after bisection.

½ A.		½ B.	
Number of pulsations.	Seconds of rest.	Number of pulsations.	Seconds of rest.
56	10	82	180
150	150	51	20
68	335	14	60
130	30	13	50
46	45	46	45
2	10	38	65
99	66	18	45
103	360	23	60
12	4	35	130
		105	70

Pauses now become longer, and swimming bouts shorter.

Twenty-four hours after the operation.

½ A.		½ B.	
Number of pulsations.	Seconds of rest.	Number of pulsations.	Seconds of rest.
2	363	50	20
12	362	81	25
4	666	37	101
25	300	2400	60

But although in the case of Sarsia the lesser endurance of the smaller segment than of the larger cannot be regarded as a general rule, it may be so regarded, as already stated, in the case the larger segments as regards the deprivation of oxygen. This is shown by the greater regularity

of Aurelia. The following experiment exemplifies this particular rule even more prettily than does the one quoted in the text, from the fact that the segments survived the operation for a greater number of days.

An Aurelia having a regular and well-sustained rhythm of twenty per minute was divided as already described in the text. In five-minute intervals on successive days the average rates of the four segments were as follows:—

Four hours after the operation.

Seg. ½.	Seg. ¼.	Seg. ⅛.	Seg. ⅛ A.
100	100	85	90

Next day.

88	90	64	58

Next day.

86	82	62	57

Next day.

59	45	24	20

Next day.

50	49	20	10

Next day.

43	33	18	4

Next day.

39	32	19	Dead.

of the rhythm manifested by the smaller than by the larger segments in the stale water, and the fact is presumably to be accounted for by the consideration that the ganglia in the smaller segments were more potent than those in the larger.

With regard, therefore, to the original point under consideration, I conclude that, although the size of the segments is doubtless one factor in determining the relative frequency of contraction, there are at least two other factors quite as important, viz. the relative potency of the lithocysts, and the length of time that elapses between performing the operation and observing the rhythm. Hence it is that in my experience I have found but very few examples of Dr. Eimer's rule.

Effects of Other Forms of Mutilation on the Rhythm.

The next point I have to dwell upon is one of some interest. If the manubrium of Aurelia, or of any other covered-eyed Medusa, be suddenly cut off at its base, the swimming motions of the umbrella

Next day.			
Seg. ½.	Seg. ¼.	Seg. ⅛.	Seg. ⅛ A.
33	7	Dead.	0

Next day.			
28	Dead.	0	0

Next day, the temperature unfortunately rose sufficiently to cause the death of the single surviving segment, which otherwise would probably have lived for one or two days longer.

immediately become accelerated. This acceleration, however, only lasts for a few minutes, when it gradually begins to decline, the rate of the rhythm becoming slower and slower, until finally it comes to rest at a rate considerably less than was previously manifested by the unmutilated animal. If a circular piece be now cut out from the centre of the umbrella, the rhythm of the latter again becomes temporarily quickened; but, as before, gradual slowing next supervenes. This slowing, however, proceeds further than in the last case, so that the rate at which the rhythm next becomes stationary is even less than before. If, now, another circular ring be cut from the central part of the umbrella— *i.e.* if the previously open ring into which this organ had been reduced by the former operation be somewhat narrowed from within—the same effects on the rhythm are again observable; and so on with every repetition of the operation, the rate of the rhythm always being quickened in the first instance, but then gradually slowing down to a point somewhat below the rate it manifested before the previous operation. It will here suffice to quote one experiment among many I have made in this connection:—

An Aurelia manifested a regular and sustained rhythm of ... 26
Immediately after removal of manubrium, rhythm rose to ... 36
Rate then gradually fell for a quarter of an hour, and became stationary at ... 20
Circular incision just including ovaries caused rhythm to rise to ... 26
After gradual fall during quarter of an hour, rhythm became stationary at ... 17

Another circular incision carried round midway between the
former one and the margin caused rhythm to rise to ... 24
Rate again gradually declined, and in a quarter of an hour was 12
Another circular incision was carried round as close to the
margin as was compatible with leaving the physiological
continuity of all the lithocysts intact. Rhythm rose to 14
Within a few minutes it fell to 6

Excepting the cases where the effects of shock are apparent, some such series of phenomena as those just recorded are always sure to ensue when a covered-eyed Medusa is mutilated in the way described, and this kind of mutilation, besides producing such marked effects on the *rate* of the rhythm, also produces an effect in impairing the *regularity* of the rhythm. In some specimens the latter effect is more marked than it is in others. The following series of observations will serve to give a good idea of this effect:—

An Aurelia manifested a regular and sustained rhythm of 36. Immediately after the removal of the manubrium, the rate of rhythm in successive minutes was as follows: 40, 39, 37, 35, 32, 30, 29, 26, 24, 18, 14 (40 seconds' pause), 16, 15, 14, 15, 16 (40 seconds' pause), 22, 20, 19, 15, 16, 17, 14, 13, 13, 15, 16, 16, 17, 18, 14, 12, 13, 11, 12, 9, 15, 16, 14, 12, 9, etc., the rhythm now continuing very irregular. An hour after the operation, the following were the number of contractions given in one-minute intervals, the observations being taken at intervals of ten minutes: 15, 15, 12, 22, 14, etc.

In this experiment, therefore, as soon as the acceleration and slowing-stages had been passed, viz. about a quarter of an hour after the operation, a great disturbance was observable in regularity of the rhythm ; for before the removal of the manubrium, the Medusa had been swimming for hours with perfect regularity.

Before concluding my description of these experiments, it may perhaps be as well to mention one other, which was designed to meet a possible objection to the inferences which, as I shall immediately argue, these experiments seem to sustain. It occurred to me as a remote possibility that the slowing and irregularity of the rhythm, which are observable about a quarter of an hour after the operations described, might be due to the deprivation of adequate nourishment suffered by the ganglia, in consequence of the escape of nutrient matter from the cut ends of the nutrient tubes. Accordingly, instead of cutting off the manubrium, I tried the effect of momentarily immersing it in hot water, and found that the subsequent disturbances of the rhythm were precisely similar to those which result from removal of the manubrium.

Now, to draw any inferences from such meagre facts as the above would be hazardous, unless we recognize that in so doing our inferences are not trustworthy. But, with this recognition, I think there will be no harm in briefly stating the deductions to which the facts, such as they are, would seem to point.

Physiologists are undecided as to the extent in which many apparently automatic actions may not really be actions of a reflex kind. Given any ganglio-muscular tissue which is rhythmically contracting, how are we to know whether the action of the ganglia is truly automatic, or sustained from time to time by stimuli proceeding from other parts of the organism? In most cases experiments cannot be conducted with reference to this question, but in the case of the Medusæ they may be so, and it was with the view of throwing light on this question that the experiments just described were made. Now in these experiments the fact is sufficiently obvious that mutilations of any part of the organism modify the rhythm of the marginal ganglia most profoundly. That this modification does not proceed from shock, would seem to be indicated by the facts that the first effect of the mutilation is to *quicken* the rhythm; that there is a sort of general proportion to be observed between the amount of tissue abstracted and the degree of slowing of the rhythm produced; and that the slowing effects continue for so long a time. All these facts seem to show that we have here something other than mere shock to deal with.

A strong suspicion, therefore, arises that the cause of the slowing of the rhythm which results from removing the manubrium, or a part of the general contractile tissue of the bell, consists in the destruction of some influence of an afferent character which had previously emanated from the parts of the organism which have been removed, and that the normal rhythm before the operation was partly due to a continuous reception, on the part of the ganglia, of this afferent or stimulating influence. In support of this view are the facts that the first effect of such an operation as we are considering is greatly to accelerate the rhythm, and that this acceleration then gradually declines through a period of about a quarter of an hour. These facts tend to support this view, because, if it is correct, they are what we might anticipate. If the manubrium, for instance, while *in situ* is continually supplying a gentle stimulus to the marginal ganglia, when it is suddenly cut off, the nerve-tracts through which this stimulating influence had previously been conveyed must be cut through; and as it is well known how irritable nerve-fibres are at their points of section, it is to be expected that the irritation caused by cutting these nerve-tracts, and probably also by the action of the sea-water on their cut extremities, would cause them to stimulate the ganglia more powerfully than they did before their mutilation. And here I may state that on several occasions, with vigorous specimens, I have observed a sudden removal of the manubrium to be followed, not merely with a quickening of the rhythm on the part of the bell, but with a violent and long-sustained spasm.

Again, as regards the other fact before us, it is obvious that as soon as the cut extremities of the nerves begin to die down, and so gradually to lose their irritability, the effect on the rhythm would be just what we observe it to be, viz. a gradual slowing till the rate falls considerably below that which was exhibited by the unmutilated animal. And even the *irregularity* which is at this stage so frequently observable is, I think, what we should expect to find if this view as to the essentially reflex character of the natural rhythm is the true one.

If this view is the true one, the question next arises as to the nature of the process which goes on in the excitable tissues, and which afterwards acts as a stimulus on the ganglionic tissues. This question, however, I am quite unable to answer. Whether the process is one of oxygenation, of chemical changes exerted by the sea-water, or a process of any other kind, further experiments may be able to show; but meanwhile I have no suggestion to offer.

Effects of lessening the Amount of Tissue adhering to a Single Ganglion.

The above experiments led me to try the effects of cutting out a single lithocyst of Aurelia, and, after the rhythm of the detached segment had become regular, progressively paring down the contractile tissues around the ganglion. I found that this process had no very marked effect on the rhythm, until the paring reached within an inch or two of the ganglion: then, however, the effect began to show itself, and with every successive paring it became more marked. This effect consisted in slowing the rate of the rhythm, but more especially in giving rise to prolonged pauses: indeed, if only a very little contractile tissue was left adhering to the ganglion, the pauses often became immensely prolonged, so that one might almost suppose the ganglion to have entirely ceased discharging. But if a stimulus of any kind were then applied, the rhythmic discharges at once recommenced. These generally continued for some little time at a slower rate than that which they had manifested before they were affected by the paring down of the contractile tissue.

Effects of Temperature on the Rhythm.

The effects of temperature on the rhythm of Medusæ are very decided. For instance, a specimen of Sarsia which in successive minutes gave the following number of pulsations, 16, 26, 0, 0, 26, gave sixty pulsations during the next minute, while a spirit-lamp was held under the water in which the Medusa was swimming. If hot water be added to that in which Sarsia are contained until the whole is about milk-warm, their swimming motions become frantic. If the same experiment be performed after the margins of the Sarsiæ have been removed, the paralyzed bells remain quite passive, while the severed margins exhibit the frantic motions just alluded to.

In the case of Aurelia aurita, the characteristic effects of temperature on rhythm may be better studied than in that of Sarsia, from the fact that the natural motions are more rhythmical and sustained in the former than in the latter genus. I have, therefore, in this connection made more observations on Aurelia than on Sarsia. The following may be taken as a typical experiment.

A small and active specimen of Aurelia contracted

with the greatest regularity 33 times per minute in water kept at 34°; but on transference to water kept at 49°, the contractions always became irregular, in respect (*a*) of not having a perfectly constant rhythm, and (*b*) of exhibiting frequent pauses, which was never the case in colder water. The rate of rhythm in the warmer water varied from 37 to 49; and as in these observations no allowance was made for the occurrence of the pauses, the actual rate of rhythm during the swimming motions was about 60 per minute. The following are some sample observations in the case of this specimen :—

Temperature of water (Fahr.)	Number of pulsations.	Seconds of rest.
40°	41	5
,,	49	4
Transferred to 34°	33	0
,, ,,	33	0
,, ,,	33	0
,, ,,	33	0
Transferred to 49°	45	4
,, ,,	39	10
,, ,,	37	15
Transferred to 34°	20	0
,, ,,	30	0
,, ,,	33	0
,, ,,	33	0
,, ,,	33	0
,, ,,	33	0

This rate continued quite regularly for a quarter of an hour, when the observation terminated.

It might naturally be supposed that when the alterations of temperature between 34° and 49° produce such marked effects on the rhythm, still greater alterations would be attended with still greater effects. Such, however, is not the case. Water at 70° or 80°, for instance, has the effect of permanently *diminishing* the rate of the rhythm, after having temporarily raised it for a few seconds. The following experiment will serve to convey a just estimation of these facts.

An Aurelia whose rhythm in water at 40° was very regular at eighteen per minute, was suddenly transferred to water at 80°. In the immediately succeeding minutes the rhythm was 22, 20, 14. The latter rate continued for nearly half an hour, when the observation terminated.

The effect of very warm water, therefore, is to slow the rhythm, as well, I may add, as to enfeeble the vigour of the contractions. The case of Medusæ thus differs, in the former respect, from that of the heart; and I think the reason of the difference is to be found in the following considerations. Even slight elevations of temperature are quickly fatal to the Medusæ, so it becomes presumable that considerable elevations act very destructively on the neuro-muscular tissues of those animals. This destructive effect of high temperatures may, therefore, very probably counteract the stimulating effect which such temperatures would otherwise exert on the natural rhythm, and hence a point would somewhere be reached at which the destructive effect would so far overcome the stimulating effect as to slow the rhythm. That this is probably the true, as it certainly is the only explanation to be rendered, will, I think, be conceded when I further state that if an Aurelia be left for some little time in water at 80°, and then again transferred to water at 30° or 40°, its original rate of rhythm at the latter temperature does not again return, but the rhythm remains permanently slowed. And, in favour of the explanation just offered, it may be further pointed out that the first effect of sudden immersion in heated water is to *quicken* the rhythm, it not being for a few seconds, or for even a minute or two after the immersion, that the rhythm becomes slowed. Lastly, the slowing takes place gradually; and this is what we should expect if, as is probable, the destructive effect takes somewhat more time to become fully developed than does the stimulating effect.

Before leaving the subject of temperature in relation to rhythm, I must say a few words on the effects of cold. The following may be regarded as typical experiments.

An Aurelia presenting a regular rhythm of twenty per minute in water at 45° was placed in water at 19°. Soon after the transference the rhythm began to slow, and the strength of the contractions to diminish. Both these phenomena rapidly became more and more pronounced, till the rhythm fell to ten per minute (still quite regular), and the contractions ceased to penetrate the muscular tissue further than an inch or so from the marginal ganglia. Shortly after this stage pauses became frequent, but mechanical or other irritation always originated a fresh swimming bout. Next, only one very feeble contraction was given at long and irregular intervals, a contraction so feeble that it was restricted to the immediate vicinity of the lithocyst in which it originated. Soon after this stage irritability towards all kinds of stimuli entirely ceased, including even strong spirit dropped on the under surface of the animal when taken momentarily out of the water. All these stages thus described were passed through rapidly, the whole series occupying rather less than five minutes. On now leaving the specimen for ten minutes and then restoring it to its original water at 45°, all the above-mentioned stages were passed through in reverse order. The first faint marginal contraction was confined to the immediate vicinity of the prepotent lithocyst, and all subsequent contractions continued to be so for the next three minutes. Rhythm very slow. Contractions now began to penetrate round the margin, and in eight minutes from the restoration had gone all the way round, the rate of their rhythm meanwhile increasing. In two minutes more all the umbrella was contracting at the rate of fifteen per minute.

In another specimen, subjected to the same conditions, the rate of recovery was even more rapid, occupying only two minutes altogether; but in every case the process of recovery is a gradual one, and differs only in the time it occupies in passing through the various stages.

Effects of Freezing Medusæ.

In conclusion, I will describe some rather interesting experiments that consisted in freezing some specimens of Aurelia into a solid block of ice. Of course, as sea-water had to be employed, the cold required was very considerable; but I succeeded in turning out the Medusæ encased on all sides in a continuous block of sea-water. By now immersing this block in warm water, I was able to release the contained specimens, which then presented a very extraordinary appearance. The thick and massive gelatinous bell of a Medusa is, as every one knows, chiefly composed of sea-water, which everywhere enters very intimately into the structure of the tissue. Now, all this sea-water was, of course, frozen *in situ*, so that the animals were everywhere and in all directions pierced through by an innumerable multitude of ice-crystals, which formed a very beautiful meshwork, pervading the whole substance of their transparent tissues.

These experiments were made in order to ascertain whether the Medusæ, after having been thus completely frozen, would survive on being again thawed out, and, if so, whether the freezing process would exert any permanent influence on the rate of their rhythm. Now in all the cases the Medusæ, after having been thawed out, presented a ragged appearance, which was due to the disintegrating effect exerted by the ice-crystals while forming in the tissues; yet notwithstanding this mechanical injury superimposed on the physiological effects of such extreme cold, all the Medusæ recovered on being restored to sea-water of the normal temperature. The time occupied by the process of recovery varied in different individuals from a few minutes to half an hour or more, and it was observable that those specimens which recovered soonest had the rate of their rhythm least affected by the freezing. In no case, however, that I observed did the rate of the rhythm after the freezing return fully to that which had been manifested before the freezing.

Effects of Certain Gases on the Rhythm.

Oxygen.—I will now conclude my remarks on rhythm by very briefly describing the effects of certain gases. Oxygen forced under pressure into sea-water containing Sarsiæ has the effect of greatly accelerating the rate of their rhythm. The following observation on a single specimen will serve to render this apparent.

Number of pulsations given by Sarsia in successive five-minute intervals.

In ordinary sea-water	472, 527, 470
In oxygenated sea-water	800
In ordinary sea-water	268, 350, 430

It will be seen from this observation that the acceleration of the rhythm due to the oxygenation was most marked; indeed, the pulsations followed one another so rapidly that it was no easy matter to count them. It must also be stated that while the animal was under the influence of oxygen, the duration of the natural pauses between the swimming bouts was greatly curtailed, the swimming motions, in fact, being almost quite continuous throughout the five minutes that the Medusa was exposed to such influence. Lastly, it will be observed from the above table that the unnatural amount of activity displayed by the organism while in the oxygenated water entailed on it a considerable degree of exhaustion, as shown by the fact that even a quarter of an hour after its restoration to normal water its original degree of energy had not quite returned.

Carbonic acid.—As might be expected, this gas has the opposite effects to those of oxygen. It is therefore needless to say more about this agent, except that if administered in large doses it destroys both spontaneity and irritability. Nevertheless, if its action is not allowed to last too long, the Medusæ will fully recover on being again restored to normal sea-water.

Nitrous oxide.—This gas at first accelerates the motions of Sarsia, but eventually retards them. I omitted, however, to push the experiment to the stage of complete anæsthesia, which would doubtless have supervened had the pressure of the gas been sufficiently great.

Deficient aëration.—It may now be stated that the Medusæ are exceedingly sensitive to such slight carbonization of the water in which they are contained as results from their being confined in a limited body of it for a few hours. The rhythm becomes slowed and the contractions feeble, while the pauses between the swimming bouts become more frequent and prolonged. If the water is not changed, all these symptoms become more marked, and, in addition, the rhythm becomes very irregular.

Eventually the swimming motions entirely cease; but almost immediately after the animals are restored to normal sea-water, they recover themselves completely, the rate and regularity of their rhythm being then quite natural. The suddenness with which this return to the normal state of things is effected cannot but strike the observer as very remarkable, and I may mention that it takes place with equal suddenness at whatever stage in the above-described process of asphyxiation the transference to normal sea-water is accomplished.

CHAPTER VIII.

ARTIFICIAL RHYTHM.

IF the umbrella of Aurelia aurita has been paralyzed by the removal of its lithocysts, and if it is then subjected to faradaic stimulation of minimal intensity, the response which it gives is not tetanic, but rhythmic. The rate of this artificial rhythm varies in different specimens, but the limits of variation are always within those which are observed by the natural rhythm of different specimens. The artificial rhythm is not in every case strictly regular; but by carefully adjusting the strength of the current, and by shifting the electrodes from one part of the tissue to another until the most appropriate part is ascertained, the artificial rhythm admits in most cases of being rendered tolerably regular, and in many cases as strictly regular as is the natural rhythm of the animal. To show this, I append a tracing of the artificial rhythm (Fig. 25), which may be taken as a fair sample of the most perfect regularity that can be obtained by minimal faradaic stimulation.*

* This and all the subsequent tracings I obtained by the method already described.

Fig. 25.

This artificial rhythm may be obtained with a portion of irritable tissue of any size, and whether a large or small piece of the tissue employed be included between the electrodes.

As the fact of this wonderfully rhythmic response to faradaic irritation was quite unexpected by me, and as it seemed to be a fact of great significance, I was led to investigate it in as many of its bearings as time permitted. First, I tried the effect on the rhythm of progressively intensifying the strength of the faradaic current. I found that with each increment of the current the rate of the rhythm was increased, and this up to the point at which the rhythm began to pass into tetanus due to summation of the successive contractions. But between the slowest rhythm obtainable by minimal stimulation and the most rapid rhythm obtainable before the appearance of tetanus, there were numerous degrees of rate to be observed. I here append another tracing, to show the effect on the rate of the rhythm of alterations in the strength of the current (Fig. 26).

It will also be observed from this tracing that, in consequence of the current having been strengthened slightly beyond the limit within which strictly rhythmic response was attainable, the curves in the middle part of the tracing, where the current was strengthened, are slightly irregular. This irregularity is, of course, due to the first appearance of tumultuous tetanus. If the faradaic stimulation had in this case been progressively made still stronger, the irregularity would have become still more pronounced up to a certain point, when it would gradually have begun to pass into more persistent tetanus. But as in this case, instead of strengthening the current still further, I again weakened it to its original intensity, the rhythm immediately returned to its original rate and regularity.

Such being the facts, the question arises as to their interpretation. At first I was naturally inclined to suppose that the artificial rhythm was due to a periodic variation in the strength of the stimulus, caused by some slight breach of contact between the terminals and the tissue on each contraction of the latter. This supposition, of course, would divest the phenomena in question of all physiological meaning, and I therefore took pains in the first instance to exclude it. This I did in two ways: first, by observing that in many cases (and especially in Cyanæa capillata) the rate of the rhythm is so slow that the contractions do not follow one another till a considerable interval of total relaxation has intervened; and second, by placing the terminals close together, so as to include only a small piece of tissue between, and then firmly pinning the tissue all round the electrodes to a piece of wood placed beneath the Medusa. In this way the small portion of tissue which served as the seat of stimulation was itself prevented from moving, and therefore the rhythmic motions which the rest of the Medusa presented cannot have been due to any variations in the quality of the contact between the electrodes and this stationary seat of stimulation.

Any such merely mechanical source of fallacy being thus, I think, excluded, we are compelled to regard the facts of artificial rhythm as of a purely physiological kind. The question, therefore, as to the explanation of these facts becomes one of the highest interest, and the hypothesis which I have framed to answer it is as follows. Every time the tissue contracts it must as a consequence suffer a certain amount of exhaustion, and therefore must become slightly less sensitive to stimulation than it was before. After a time, however, the exhaustion will pass away, and the original degree of sensitiveness will thereupon return. Now, the intensity of faradaic stimulation which is alone capable of producing rhythmic response, is either minimal or but slightly more than minimal in relation to the sensitiveness of the tissue when fresh; consequently, when this sensitiveness is somewhat lowered by temporary exhaustion, the intensity of the stimulation becomes somewhat less than minimal in relation to this lower degree of sensitiveness. The tissue, therefore, fails to perceive the presence of the stimulus, and consequently fails to respond. But so soon as the exhaustion is completely recovered from, so soon will the tissue again perceive the presence of the stimulus; it will therefore again respond, again become temporarily exhausted, again fail to perceive the presence of the stimulus, and again become temporarily quiescent. Now it is obvious that if this process occurs once, it may occur an indefinite number of times; and as the conditions of nutrition, as well as those of stimulation, remain constant, it is manifest that the responses may thus become periodic.

In order to test the truth of this hypothesis, I made the following experiments. Having first noted the rate of the rhythm under faradaic stimulation of minimal intensity, without shifting the electrodes or altering the intensity of the current, I discarded the faradaic stimulation, and substituted for it single induction shocks thrown in with a key. I found, as I had hoped, that the *maximum* number of these single shocks which I could thus throw in in a given time *so as to procure a response to every shock*, corresponded with the number of contractions which the tissue had previously given during a similar interval of time when under the influence of the faradaic current of similar intensity. To make this quite clear, I shall describe the whole course of one such experiment. The deganglionated tissue under the influence of minimal faradaic stimulation manifested a perfectly regular rhythm of thirty contractions per minute, or one contraction in every two seconds. While the position of the platinum electrodes and the intensity of the current remained unchanged, single induction shocks were now administered with a key at any intervals which might be desired. It was found that if these single induction stimuli were administered at regular intervals of two seconds or more, the tissue responded to every stimulus; while if the stimuli were thrown in more rapidly than this, the tissue did not respond to every stimulus, but only to those that were separated from one another by an interval of at least two seconds' duration. Thus, for instance, if the shocks were thrown in at the rate of one a second, the tissue only, but always, responded to every alternate shock. And similarly, as just stated, if any number of shocks were thrown in, the tissue only responded once in every two seconds. Now, as this rate of response precisely coincided with the rate of rhythm previously shown by the same tissue under the influence of faradaic stimulation of the same intensity, the experiment tended to verify the hypothesis which it was designed to test.

I may give one other experiment having the same object and tendency. Employing single induction shocks of slightly more than minimal intensity, and throwing them in at twice the rate that was required to produce a strong response to every shock, I found that midway between every two strong responses there was a weak response. In other words, a stimulus of uniform intensity gives rise alternately to a strong and to a weak contraction, as shown in the appended tracing (Fig. 27). It will be observed that in this tracing each large curve represents the whole time occupied by the strong contraction, the latter beginning at the highest point of the curve on the left-hand side in each case. The effect of the weak contraction is that of momentarily interrupting the even sweep of diastole after the strong contraction, and therefore the result on the tracing is a slight depression in the otherwise even curve of ascent. Lest any doubt should arise from the smallness of the curves representing the weak contractions that the former

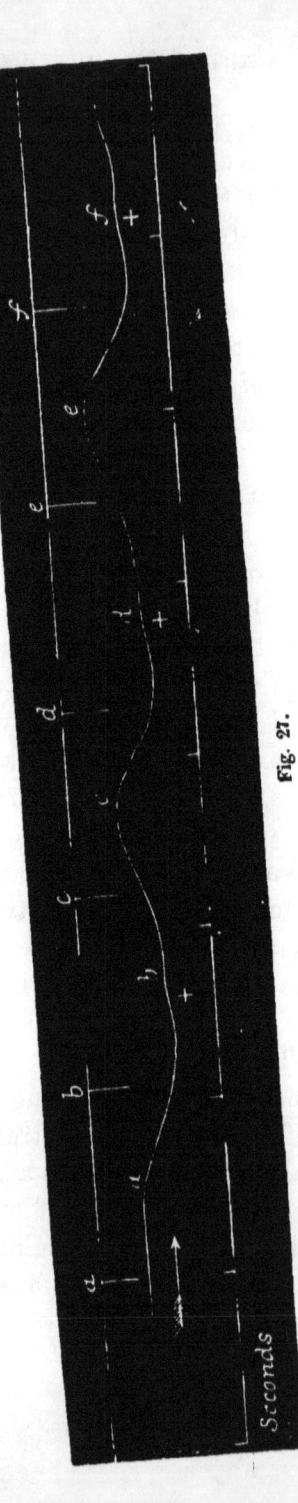

Fig. 27.

are in some way accidental, I may draw attention to the fact that the period of latent stimulation is the same in the case of all the curves. To render this apparent, I have placed crosses below the smaller curves, which show in each case the exact point where the depressing effect of these smaller curves on the ascending sweeps of the larger curves first become apparent—*i.e.* the point at which the feeble contraction begins. Now, what I wish to be gathered from the whole tracing is this. If the strength of the induction shocks had been much greater than it was, *all* the contractions would have become strong contractions, and tetanus would have been the result. But, as the strength of the induction shocks was only slightly more than minimal, the exhaustion consequent on every strong contraction so far diminished the irritability of the tissue that when, during the process of relaxation, another shock *of the same intensity* was thrown in, the stimulus was only strong enough, in relation to the diminished irritability of the partly recovered tissue, to cause a feeble contraction. And these facts tend still further to substantiate the hypothesis whereby I have sought to explain the phenomena of artificial rhythm.

Now, I think that the strictly rhythmic action of the paralyzed swimming-bell of Aurelia in answer to constant stimulation is a fact of the highest significance; for here we have a tissue wholly, or almost wholly, deprived of its centres of spontaneity, yet pulsating as rhythmically in answer to artificial stimulation as it previously did in answer to ganglionic stimulation.* Does not this tend to show that for the production of the natural rhythm the presence of the ganglionic element is non-essential; that if we merely suppose the function of this element to be that of supplying a constant stimulus of a low intensity, without in addition supposing the presence of any special resistance-mechanism to regulate the discharges, the periodic sequence of systole and diastole would assuredly result; and, therefore, that the rhythmical character of the natural swimming motions is dependent, not on the peculiar relations of the ganglionic, but on the primary qualities of the contractile tissue? Or, if we do not go so far as this (and, as I may parenthetically observe, I am not myself inclined to go so far), must we not at least conclude that the natural rhythm of these tissues is not *exclusively* due to any mechanism whereby the discharges of the ganglia are interrupted at regular intervals; but that whether these discharges are supposed to be interrupted or continuous, the natural rhythm is probably in a large measure due to the same cause as the artificial rhythm, viz. in accordance with our previous hypothesis, to the alternate exhaustion and recovery of the excitable tissues? This much, at least, must be allowed even by the most cautious

---

* It will not be forgotten that there are a multitude of ganglion-cells distributed throughout the contractile tissues of the Medusæ; but forasmuch as these are comparatively rarely instrumental in originating stimulation, I think it is probable that artificial stimulation acts directly on the contractile tissues, and not through the medium of these scattered cells.

of critics, viz. that if, as current views respecting the theory of rhythm would suppose, it is exclusively the ganglionic element which in the unmutilated Aurelia causes the rhythm of the swimming motions by intermittent stimulation, surely it becomes a most unexpected and unaccountable fact, that after the removal of this element the contractile tissues should still persist in their display of rhythm under the influence of constant stimulation. At any rate no one, I think, will dispute that the facts which I have adduced justify us in reconsidering the whole theory of rhythm as due to ganglia.

As I have already said, I am not inclined to deny that there is probably some truth in the current theory of rhythm as due to ganglia; I merely wish to point out distinctly that this theory is inadequate, and that in order to cover all the facts it will require to be supplemented by the theory which I now propose. The current theory of rhythm as due to ganglia attributes the whole of the effect to the ganglionic element, and thus fails to meet the fact of a rhythm which is artificially produced after the ganglionic element has been removed. It also fails to meet a number of other facts of the first importance; for it is beyond all doubt that rhythmic action of the strictest kind occurs in an innumerable multitude of cases where it is quite impossible to suppose anything resembling ganglia to be present. Not to mention such cases as the Snail's heart, where the most careful scrutiny has failed to detect the least vestige of ganglia, but to descend at once to the lowest forms of animal and vegetable life, rhythmic action may here be said to be the rule rather than the exception. The beautifully regular motions observable in some Algæ, Diatomaceæ, and Ocillatoriæ, in countless numbers of Infusoria, Antherozoids, and Spermatozoa, in ciliary action, and even in the petioles of Hedysarum gyrano, are all instances (to which many others might be added) of rhythmical action where the presence of ganglia is out of the question. Again, in a general way, is it not just as we recede from these primitive forms of contractile tissue that we find rhythmic action to become less usual? And, if this is so, may it not be that those contractile tissues which in the higher animals manifest rhythmic action are the contractile tissues which have longest retained their primitive endowment of rhythmicality? To my mind it seems hard to decide in what respect the beating of a Snail's heart differs from that of the pulsatile vesicles of the Infusoria; and I do not think it would be much easier to decide in what essential respect it differs from the beating of the Mammalian heart. The mere fact that the presence of ganglia can be proved in the one case and not in the other, seems to me scarcely to justify the conclusion that the rhythm is in the one case wholly dependent, and in the other as wholly independent, of the ganglia. At any rate, this fact, if it is a fact, is not of so self-evident a character as to recommend to us the current theory of ganglionic action on *à priori* grounds.

Coming, then, to experimental tests, we have already seen that in the deganglionated swimming organ of Aurelia aurita, rhythmic response is yielded to constant faradaic stimulation of low intensity. The next question, therefore, which presents itself in relation to our subject is as to whether other modes of constant stimulation elicit a similar response. Now, in a general way, I may say that such is the case, although I have chosen faradaic stimulation for special mention, because, in the first place, its effect in producing rhythmic action is the most certain and precise; and, in the next place, the effects of administering instantaneous shocks at given intervals admit of being compared with the effects of constant faradaic stimulation better than with any other kind of constant stimulation. Nevertheless, as just stated, other modes of constant stimulation certainly have a more or less marked effect in producing rhythmic response. The constant current, during the whole time of its passage, frequently has this effect in the case of the paralyzed nectocalyx of Sarsia; and dilute spirit, or other irritant, when dropped on the para'yzed swimming organ of Aurelia aurita, often gives rise to a whole series of rhythmical pulsations, the systoles and diastoles following one another at about the same rate as is observable in the normal swimming motions of the unmutilated animal.

From this it will be seen that, both in the case of mechanical and of chemical stimulation, the same tendency to the production of rhythmic response on the part of the paralyzed tissues of Aurelia may be observed as in the case of electrical stimulation. The principal differences consist in the rhythm being much less sustained in the former than in the latter case. But, by experimenting on other species of Medusæ, I have been able to obtain, in response to mechanical and chemical stimulation, artificial rhythm of a much more sustained character than that which, under such modes of stimulation, occurs in Aurelia. I have no explanation to offer why it is that some species, or some tissues, present so much more readiness to manifest sustained rhythm under certain modes of stimulation, and less readiness to manifest it under other modes, than do other species or tissues. Probably these differences depend on some peculiarities in the irritability of the tissues which it is hopeless to ascertain; but, in any case, the facts remain, that while Aurelia, Cyanæa, and the covered-eyed Medusæ generally are the best species for obtaining artificial rhythm under the influence of faradaic stimulation, some of the naked-eyed Medusæ are the best species for obtaining it under the influence of the constant current, and also under that of mechanical and chemical stimulation. I have already spoken of this effect of the constant current in the case of Sarsia; I shall now proceed to describe the effects of mechanical and chemical stimulation on the same species.

It is but rarely that artificial rhythm can be produced in the paralyzed nectocalyx of Sarsia by means of mechanical stimulation, but in the case of the manubrium, a very decided, peculiar, and persistent rhythm admits of being produced by this means. In this particular species, the manubrium never exhibits any spontaneous motion after the ganglia of the nectocalyx have been removed. But if it be nipped with the forceps, or otherwise irritated, it contracts strongly and suddenly; it then very slowly and gradually relaxes until it has regained its original length. After a considerable interval, and without the application of any additional stimulus, it gives another single, sudden, though slight contraction, to be again followed by gradual relaxation and a prolonged interval of repose, which is followed in turn by another contraction, and so on. These sudden and well-marked contractions occur at intervals of many seconds, and show a decided tendency to rhythmic periodicity, though the rhythm is not always perfectly exact. This intensely slow rhythm, as the result of injury, may continue for a long time, particularly if the injury has been of a severe character. There can be no doubt, therefore, that the mechanical (or other) injury in this case acts as a source of constant irritation; so that here again we have evidence of rhythmic action independent of ganglia, and caused by the alternate exhaustion and recovery of contractile tissues.*

* We may pretty safely conclude that ganglia are altogether absent in the manubrium of Sarsia, not only because Schultz has failed to detect them in this organ microscopically, but also because of the complete absence of spontaneity which it manifests. I may here mention that this case of the manubrium of Sarsia is precisely analogous to another which I have observed in a widely different tissue, namely, the tongue of the frog. Here,

With regard to artificial rhythm caused by chemical stimuli, by far the most conspicuous instance that I have observed is that of the paralyzed nectocalyx of Sarsia. This consists in a highly peculiar motion of a flurried, shivering character, which is manifested by this organ when its marginal ganglia have been removed and it is exposed to the influence of faintly acidulated water. Now, when read in the light of the foregoing facts, there can be no doubt that the present one falls into its place very satisfactorily: it is an additional and very valuable instance of the display of artificial rhythm under the influence of a constant stimulus of low intensity; for the shivering motions of the mutilated nectocalyx under these circumstances are most unmistakably of a rhythmic nature. Viewed from a little distance, indeed, these motions are not distinguishable from the natural swimming motions of the unmutilated animal, except that, not being of quite such a powerful character, they are not so effective for locomotion. Viewed more closely, however, it may frequently be seen that the whole bell does not contract simultaneously, but that, as it were, clouds of contraction pass now over one part and now over another. Still, whether the contractions are partial or universal, they are more or less rhythmical. As this was the only case that had ever been observed of

too, the presence of ganglion-cells has never been observed microscopically, though specially sought for by Dr. Klein and others. Yet, under the influence of mechanical and other modes of stimulation, I find that I am able to make the excised organ pulsate as rhythmically as a heart.

rhythm as due to a constant chemical stimulus, I studied it with much care, and shall now give a full description of what appears to me an important body of physiological facts.

Ten to twenty drops of acetic acid having been added to one thousand cubic centimetres of sea-water, and the paralyzed bell of Sarsia having been placed in the mixture, an interval of about half a minute will elapse before any movement begins. Sooner or later, however, the artificial rhythm is sure to be induced, and it will then continue for a variable time—occasionally as long as an hour, and generally for a considerable number of minutes. After it ceases it may often be made to recommence, either by adding a few more drops of acid to the sea-water, or by supplying an additional stimulus to the bell by nipping it with the forceps. Eventually, however, all movement ceases, owing to the destruction of irritability by the action of the acid. By this time the whole inner surface of the bell has become strongly opalescent, owing to the destructive influence of the acid on the epithelial cells which overspread the irritable tissues. The latter fact is worth mentioning, because in no case does the artificial rhythm set in until this opalescence has begun to show itself; and as this opalescence is but an optical expression of the damage which the epithelial coat is undergoing, the explanation of the time which elapses after the first immersion of the bell in the acidulated water and the commencement of the artificial rhythm no doubt is, that during this time the acid has not obtained sufficient access to the excitable tissues to serve as an adequate stimulus.

During the soaking stage of the experiment—*i.e.* before the artificial rhythm begins—the excitability of the tissues may be observed progressively and abnormally to increase; for soon after the soaking stage begins, in response to a single nip with the forceps the bell may give two or three locomotor contractions, instead of a single one, as is *invariably* the case with a paralyzed bell of Sarsia in normal water. Later on during the soaking stage, four or five successive contractions may be yielded in response to a single mechanical stimulus, and shortly after this a whole bout of rhythmic contractions may be started by the same means. Indeed, in some cases the artificial rhythm in acidulated water requires such a single additional stimulus for its inauguration, the shivering movements failing to begin spontaneously, but beginning immediately upon the application of the additional stimulus. Similarly, after the shivering movements have ceased, a fresh bout may very often be started by again giving the motionless nectocalyx a single stimulation. The interpretation of these facts would seem to be that the general irritability of the excitable tissues is exalted by the universal and constant stimulus supplied by the acid to an extent that is just bordering on that which gives rise to rhythmic movement, so that when the violent contraction is given in response to the mechanical stimulus, the disturbance serves to start the rhythmic movement.

If a paralyzed nectocalyx, while manifesting its artificial rhythm in acidulated sea-water, be suddenly transferred to normal sea-water, the movements do not cease immediately, but continue for a considerable time. This fact can easily be explained by the very probable, and indeed almost necessary, supposition that it takes some time after the transference to the normal sea-water for the acid to be washed out from contact with the excitable tissues. Sooner or later, however, as we should expect, in the normal sea-water the rhythmic movements entirely cease, and the bell becomes quiescent, with a normal irritability as regards single stimuli. If it be now again transferred to the acidulated water, after a short interval the rhythmic movements will again commence, and so on during several repetitions of this experiment, until the irritability of the tissues has finally become destroyed by the influence of the acid.

Other chemical irritants which I have tried produce substantially similar effects on the paralyzed bell of Sarsia. I shall, therefore, only wait to describe the influence of one of these irritants, the action of which in some respects differs from that of acids, and which I have found to be one of the most unfailing in its power to produce the rhythmic movements in question. This irritant is glycerine, and in order to produce its full effect it requires to be added to the sea-water in about the proportion of five per cent. The manifestation of artificial rhythm in solutions of this kind is quite unfailing. It begins after an exposure of from fifteen to thirty seconds, and continues for a variable number of seconds. It generally begins with powerful contractions, of a less shivering character than those which are produced by acids, and therefore still more closely resembling the normal swimming motions of the unmutilated animal. Sometimes, however, the first manifestation of the artificial rhythm is in the form of a few gentle rhythmic contractions, to be followed by a few seconds of quiescence, and then by the commencement of the sustained bout of strong contractions. In either case, when the bout of strong contractions sets in, the rate of the rhythm becomes progressively and rapidly increased, until it runs up into incipient tetanus. The rate of the rhythm still quickening, the tetanus rapidly becomes more and more pronounced, till at last the bell becomes quiescent in tonic spasm.*

If the bell is still left in the glycerine solution nothing further happens; the tissues die in this condition of strong systole. But if the bell be transferred to normal sea-water immediately after, or, still better, slightly before the tonic spasm has become complete, an interesting series of phenomena is presented. The spasm persists for a long

* Sometimes, however, the order of events is slightly different, the advent of the spasm being more sudden, and followed by a mitigation of its severity, the bell then exhibiting what is more usually the first phase of the series, namely, the occurrence of the locomotor-like contractions. Occasionally, also, rhythmical shivering contractions may be seen superimposed on the general tonic contraction, either in a part or over the whole of the contractile tissues.

time after the transference without undergoing any change, the length of this time depending on the stage in the severity and the spasm at which the transference is made. After this time is passed, the spasm becomes less pronounced than it was at the moment of transference, and a reversion takes place to the rhythmic contractions. The spasm may next pass off entirely, leaving only the rhythmic contractions behind. Eventually these too fade away into quiescence, but it is remarkable that they leave behind them a much more persistent exaltation of irritability than is the case with acid. For in the case of glycerine, the paralyzed bell which has been exposed to the influence of the irritant and afterwards become quiescent in normal sea-water, will often continue for hours to respond to single stimuli with a bout of rhythmic contractions. This effect of glycerine in producing an extreme condition of exalted irritability is also rendered apparent in another way; for if, during the soaking stage of the experiment—*i.e.* before the first of the rhythmic contractions has occurred— the bell be nipped with the forceps, the effect may be that of so precipitating events that the whole of the rhythmic stages are omitted, and the previously quiescent bell enters at once into a state of rigid tonic spasm. This effect is particularly liable to occur after prolonged soaking in weak solutions of glycerine.

As in the case of stimulation by acid, so in that of stimulation by glycerine, the artificial rhythm never begins in any strength of solution until the epithelial surface has become opalescent to a considerable degree.

In the case of stimulation by glycerine, the behaviour of the manubrium is more unequivocal than it is in the case of stimulation by acid. I have therefore reserved till now my description of the behaviour of this organ under the influence of constant chemical stimulation. This behaviour is of a very marked though simple character. The manubrium is always the first to respond to the stimulation, its retraction preceding the first movements of the bell by an interval of several seconds, so that by the time the bell begins its rhythmic response the manubrium is usually retracted to its utmost. The initial response of the manubrium is also rhythmic, and the rhythm which it manifests—especially if the glycerine solution be not over-strong—is of the same slow character which has already been described as manifested by this organ when under the influence of mechanical stimulation. The rhythm, however, is decidedly quicker in the former than in the latter case.

Lastly, with regard to the marginal ganglia, it is to be observed that in the case of all the chemical irritants I have tried, if unmutilated specimens of Sarsia be immersed in a sea-water solution of the irritant which is of a sufficient strength to evoke artificial rhythm in paralyzed specimens, the spontaneity of the ganglia is destroyed in a few seconds after the immersion of the animals, *i.e.* in a shorter time than is required for the first appearance of artificial rhythm. Consequently, whether the specimens experimented upon be entire or paralyzed by removal of their margins, the phenomena of artificial rhythm under the influence of chemical stimulation are the same. But although the spontaneity of the ganglia disappears before the artificial rhythm sets in, such is not the case with the reflex activity of the ganglia; for on nipping a tentacle of the quiescent bell before the artificial rhythm has set in, the bell will give a single normal response to the stimulation.

Hence, in historical order, on dropping an unmutilated specimen of Sarsia into a solution of glycerine of the strength named, the usual succession of events to be observed is as follows. First, increased activity of the normal swimming motions, to be quickly followed by a rapid and progressive decrease of such activity, till in about fifteen seconds after the immersion total quiescence supervenes. Four or five seconds later the manubrium begins to retract by rhythmical twitches, the rate of this rhythm rapidly increasing until it ends in tonic contraction. When the manubrium has just become fully retracted—or very often a little earlier—the bell suddenly begins its forcible and well-pronounced rhythmic contractions, which rapidly increase in their rate of rhythm until they coalesce into a vigorous and persistent spasm. If the animal be now restored to normal sea-water, spontaneity will return in a feeble manner; but there is always afterwards a great tendency displayed by the bell to exhibit shivering spasms instead of norma' swimming movements in response to natural or ganglionic stimulation. And, as already observed, this peculiarity of the excitable tissues is also well marked in the case of the artificial stimulation of deganglionated specimens under otherwise similar conditions.

One further experiment may here be mentioned. Having split open the paralyzed bell of Sarsia along the whole of one side from base to apex of the cone, I suspended the now sheet-like mass of tissue by one corner in the air, leaving the rest of the sheet to hang vertically downwards. By means of a rack-work support I now lowered the sheet of tissue, till one portion of it dipped into a beaker filled with a solution of glycerine of appropriate strength. After allowing this portion to soak in the solution of glycerine until it became slightly opalescent, I dropped the entire mutilated bell, or sheet of tissue, into another beaker containing sea-water. If the exposure to the glycerine solution had been of sufficient duration, I invariably found that in the normal sea-water the rhythmic movements were performed by the whole tissue-mass quite as efficiently as was the case in my other experiments, where the whole tissue-mass, and not merely a portion, had been submitted to the influence of the irritant. But on now suddenly snipping off the opalescent portion of the tissue-mass, i.e. the portion which had been previously alone submitted to the influence of the irritant, all movement in the remainder of the tissue-mass instantly ceased. This experiment I performed repeatedly, sometimes exposing a large and sometimes a small portion of the tissue to the influence of the irritant. As I invariably obtained the same result, there can be no doubt that in the case of chemical stimulation the artificial rhythm depends for its manifestation on the presence of a constant stimulus, and is not merely some kind of obscure fluttering motion which, having been started by a stimulus, is afterwards kept up independently of any stimulus.

Such being the case, I naturally expected that if I were to supply a constant stimulus of a thermal kind, I should also obtain the phenomena of artificial rhythm. In this, however, my expectations have not been realized. With no species of Medusa have I been able to obtain the slightest indication of artificial rhythm by immersing the paralyzed animals in heated water. I can only explain this fact by supposing that the stimulus which is supplied by the heated medium is of too uniform a character over the whole extent of the excitable tissues; it would seem that in order to produce artificial rhythm there must be a differential intensity of stimulation in different parts of the responding tissue, for no doubt even the excitatory influence of acidulated water is not of nearly so uniform an intensity over the whole of the tissue-area as is that of heated water.

In now quitting the subject of artificial rhythm as it is manifested by the paralyzed bell of Sarsia, it is desirable again to observe that sustained artificial rhythm cannot be produced by means of chemical irritation in the case of any one of the species of covered-eyed Medusæ that I have met with. In order to evoke any response at all, stronger solutions of the irritants require to be employed in the case of the covered than in that of the naked-eyed Medusæ, and when the responses do occur they are not of so suggestive a character as those which I thought it worth while so fully to describe. Nevertheless, even in the covered-eyed Medusæ well marked, though comparatively brief, displays of artificial rhythm may often be observed as the result of constant chemical stimulation. Thus, for instance, in the case of Aurelia, if the paralyzed umbrella be immersed in a solution of glycerine (ten to twenty per cent.), a few rhythmic pulsations of normal rate are usually given; but shortly after these pulsations occur, the tissue begins to go into a tetanus, which progressively and rapidly becomes more and more pronounced until it ends in violent tonic spasm. So that the history of events really resembles that of Sarsia under similar circumstances, except that the stage of artificial rhythm which inaugurates the spasm is of a character comparatively less pronounced.

Thus far, then, I have detailed all the facts which I have been able to collect with reference to the phenomena of artificial rhythm, as produced by different kinds of constant stimulation. It will not be forgotten that the interest attaching to these facts arises from the bearing which they have on the theory of natural rhythm. My belief is that hitherto the theory of rhythm as due to ganglia has attributed far too much importance to the ganglionic as distinguished from the contractile tissues, and I have founded this belief principally on the facts which have now been stated, and which certainly prove at least this much : that after the removal of the centres of spontaneity, the contractile tissues of the Medusæ display a marked and persistent tendency to break into rhythmic action whenever they are supplied with a constant stimulus of feeble intensity. Without waiting again to indicate how this fact tends to suggest that the natural rhythm of the unmutilated organisms is probably in large part due to that alternate process of exhaustion and restoration of excitability on the part of the contractile tissues, whereby alone the phenomena of artificial rhythm can be explained,* I shall go on to describe some further experiments which were designed to test the question whether the influences which affect the character of the natural rhythm likewise, and in the same manner, affect the character of the artificial rhythm. I took the trouble to perform these experiments, because I felt that if they should result in answering this question in the affirmative, they would tend still further to

* It is of importance to point out the fact that some of my previously stated experiments appear conclusively to prove that the natural stimulation which is supplied by the marginal ganglia of the Medusæ resembles all the modes of artificial stimulation which are competent to produce artificial rhythm in one important particular; the *intensity* of the stimulation which the marginal ganglia supply is shown by these experiments to be about the same as that which is required to produce artificial rhythm in the case of artificial stimulation. In proof of this point, I may allude particularly to the observations which are detailed on pp. 134-136.

substantiate the view I am endeavouring to uphold, viz. that the natural rhythm may be a function of the contractile as distinguished from the ganglionic tissue. Of the modifying causes in question, the first that I tried was temperature.

Having already treated of the effects of temperature on the natural rhythm, it will now be sufficient to say that we have seen these effects to be similar to those which temperature exerts on the rhythm of ganglionic tissues in general. Now, I find that temperature exerts precisely the same influence on the artificial rhythm of deganglionated tissue as it does on the natural rhythm of the unmutilated animal. To economize space, I shall only quote one of my observations in a table which explains itself. I also append tracings of another observation, to render the difference in the rate of the artificial rhythm more apparent to the eye (Fig. 28).

Temperature of water (Fahr.).	Number of contractions per minute.
25°	24
45°	40
75°	60

During the whole progress of such experiments the faradaic stimulation was, of course, kept of uniform intensity; so that the progressive acceleration is undoubtedly due to the increase of temperature alone. With each increment of temperature the rate of the artificial rhythm increases suddenly, just as it does in the case of the natural rhythm. Moreover, there seems to be a sort of rough correspondence between the amount of in-

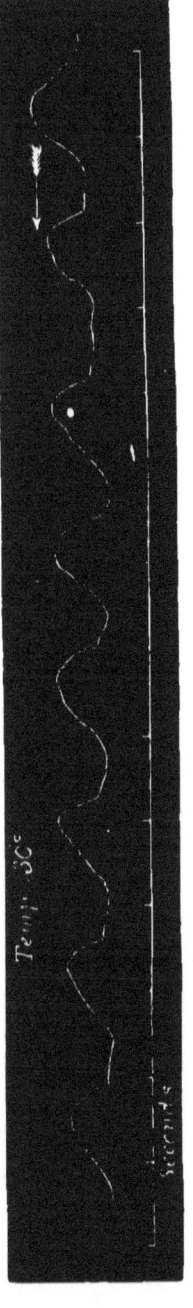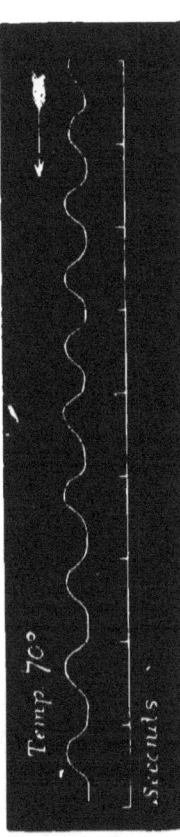

Fig. 28.

fluence that any given degree of temperature exerts on the rate of the natural and of the artificial rhythm respectively. Further, it will be remembered that in warm water the natural rhythm, besides being quicker, is not so regular as it is in cold water; thus also it is with the artificial rhythm. Again, water below 20° or above 85° suspends the natural rhythm, *i.e.* stops the contractions; and the artificial rhythm is suspended at about the same degrees. Lastly, just as there are considerable individual variations in the extent to which the natural rhythm is affected by temperature, so the artificial rhythm is in some cases more influenced by this cause than in others, though in all cases it further resembles the natural rhythm in showing some considerable degree of modification under such influence.

On the whole, then, it would be impossible to imagine two cases more completely parallel than are these of the effects of temperature on natural and on artificial rhythm respectively; and as it must be considered in the last degree improbable that all these coincidences are accidental, I conclude that the effects of temperature on the natural rhythm of Medusæ (and so, in all probability, on the natural rhythm of other ganglio-muscular tissues) are for the most part exerted, not on the ganglionic, but on the contractile element.

In order to test the effects of gases on the artificial rhythm, I took a severed quadrant of Aurelia, and floated it in sea-water, with its muscular surface just above the level of the water.

Over the tissue I lowered an inverted beaker filled with the gas the effects of which I desired to ascertain, and by progressively forcing the rim of the beaker into the water I could submit the tissue to various pressures of the atmosphere of the gas I was using. By an appropriate arrangement the electrodes passed into the interior of the beaker, and could then be manipulated from the outside, so as to be properly adjusted on the tissue. In this way I was able to observe that different gases exerted a marked influence on the rate of the artificial rhythm.

The following table gives the ratios in the case of one experiment:—

Rate of artificial rhythm, in air.	In oxygen.	In carbonic acid.
36 per minute.	50 per minute.	25 per minute.

It may here be observed that to produce these results, both carbonic acid and oxygen must be considerably diluted with air, for otherwise they have the effect of instantaneously inhibiting all response, even to the strongest stimulation. When this is the case, however, irritability returns very soon after the tissue is again exposed to air or to ordinary sea-water. But I desire it to be understood that the results of my experiments on the influence of oxygen, both on the natural and on the artificial rhythm, have proved singularly equivocal; so that as far as this gas is concerned further observations are required before the above results can be accepted as certain.

I have still one other observation of a very

interesting character to describe, which is closely connected with the current views respecting ganglionic action, and may therefore be more conveniently considered here than in any other part of this treatise. I have already stated that in no case is the manubrium of a Medusa affected as to its movements by removal of the periphery of the swimming-bell; but in the case of Sarsia a very interesting change occurs in the manubrium soon after the nectocalyx has been paralyzed by excision of its margin. Unlike the manubriums of most of the other Medusæ, this organ, in the case of Sarsia, is very highly retractile. In fresh and lively specimens the appendage in question is carried in its retracted state; but when the animals become less vigorous—from the warmth or impurity of the water in which they are confined, or from any other cause—their manubriums usually become relaxed. The relaxation may show itself in various degrees in different specimens subjected to the same conditions, but in no case is the degree of relaxation so remarkable as that which may be caused by removing the periphery of the nectocalyx. For the purpose of showing this effect, it does not signify in what condition as to vigour, etc., the specimen chosen happens to be in; for whether the manubrium prior to the operation be contracted or partially relaxed, within half an hour after the operation it is sure to become lengthened to a considerable extent.

In order to show the surprising degree to which this relaxation may proceed, I insert a sketch of a specimen both before and after the operation. The sketches are of life size, and drawn to accurate measurement (Figs. 29 and 30).*

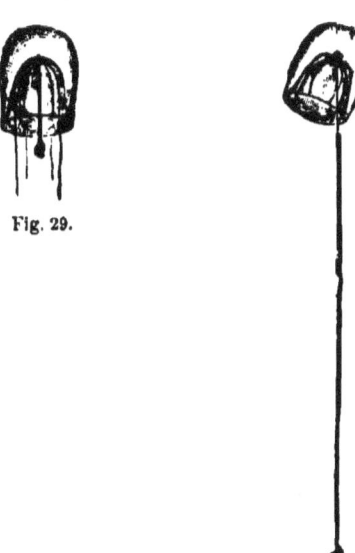

Fig. 29.

30.

With regard to this remarkable effect on the manubrium of removing the margin of the nectocalyx, it is now to be observed that in it we appear to have very unexceptionable evidence of such a relation subsisting between the ganglia of the nectocalyx and the muscular fibres of the manubrium as elsewhere gives rise to what is known as muscular tonus. This interpretation

* I may here mention that the fact of the manubrium of Sarsia undergoing this extreme elongation after the removal of the marginal ganglia, serves to render the artificial rhythm of the organ under the influence of injury, as previously described, all the more conspicuous.

of the facts cannot, I think, be disputed; and it fully explains why, in the unmutilated animal, the degree of elongation on the part of the manubrium usually exhibits an inverse proportion to the degree of locomotor activity displayed by the bell. I may here state that I have also observed indications of muscular tonus in some of the other Medusæ, but for the sake of brevity I shall now restrict myself to the consideration of this one case.

To my mind, then, it is an interesting fact that ganglionic tissue, where it can first be shown to occur in the animal kingdom, has for one of its functions the maintenance of muscular tonus; but it is not on this account that I now wish to draw prominent attention to the fact before us. Physiologists are almost unanimous in regarding muscular tonus as a kind of gentle tetanus due to a persistent ganglionic stimulation, and against this opinion it seems impossible to urge any valid objection. But, in accordance with the accepted theory of ganglionic action, physiologists further suppose that the only reason why some muscles are thrown into a state of tonus by ganglionic stimulation, while other muscles are thrown into a state of rhythmic action by the same means, is because the resistance to the passage of the stimulation from the ganglion to the muscle is less in the former than in the latter case. Here, be it remembered, we are in the domain of pure speculation: there is no experimental evidence to show that such a state of differential resistance as the theory requires actually obtains. Hence we are quite at liberty to suppose any other kind of difference to obtain, either to the exclusion of this one or in company with it. Such a supposition I now wish to suggest, and it is this—that all rhythmical action being regarded as due (at any rate in large part) to the alternate exhaustion and restoration of excitability on the part of contractile tissues, the reason why continuous ganglionic stimulation produces incipient tetanus in the case of some muscles and rhythmic action in the case of others, is either wholly or partly because the irritability of the muscles in relation to the intensity of the stimulation is greater in the former than in the latter case. If this supposition as to differential irritability be granted, my experiments on paralyzed Aurelia prove that tetanus would result in the one case and rhythmic action in the other. For it will be remembered that in these experiments, if the continuous faradaic stimulation were of somewhat more than minimal intensity, tetanus was the result; while if such stimulation were but of minimal intensity, the result was rhythmic action. Now, that in the particular case of Sarsia the irritability of the tonically contracting manubrium is higher than that of the rhythmically contracting bell is a matter, not of supposition, but of observable fact; for not only is the manubrium more irritable than the bell in response to direct stimulation of its own substance, but it is generally more so even when the stimuli are applied anywhere over the excitable tissues of the bell. And from this it is evident that the phenomena of muscular tonus, as they occur in Sarsia, tend more in favour of the exhaustion than of the resistance theory.*

I will now sum up this rather lengthy discussion. The two theories of ganglionic action may be stated antithetically thus: in both theories the accumulation of energy by ganglia is supposed to be a continuous process; but while the resistance theory supposes the rhythm to be exclusively due to an intermittent and periodic discharge of this accumulated energy on the part of the ganglionic tissues, the exhaustion theory supposes that the rhythm is largely due to a periodic process of exhaustion and recovery on the part of the respond-

---

* The evidence, however, is not altogether exclusive of the resistance theory, for it is quite possible that in addition to the high irritability of the manubrium there may be conductile lines of low resistance connecting this organ with the marginal ganglia. I entertain this supposition because, as explained in my Royal Society papers, I see reason to believe that the natural swimming movements of Sarsia are probably in part due to an intermittent discharge of the ganglia. I think, therefore, that in this particular case the ganglia supply a tolerably constant stimulation to the manubrium, while it is only at intervals that their energy overflows into the bell, and that the higher degree of irritability on the part of the manubrium ensures the tonic response of this organ at a small cost of nervous energy. How far the rhythm of the nectocalyx is to be attributed to the resistance mechanism of the ganglia, and how far to the alternate exhaustion and recovery of the contractile tissues, I think it is impossible to determine, seeing that it is impossible exactly to imitate the natural ganglionic stimulation by artificial means. But it is, I think, of importance to have ascertained at least this much, that in Sarsia the tonus of one organ and the rhythm of another, which apparently both received their stimulation from the same ganglia, must at any rate in part be attributed to a differential irritability of these organs, as distinguished from their differential stimulation.

ing tissues. Now, I submit that my experiments have proved the former of these two theories inadequate to explain all the phenomena of rhythm as it occurs in the Medusæ; for these experiments have shown that even after the removal of the only ganglia which serve as centres of natural stimulation, the excitable tissues still continue to manifest a very perfect rhythm under the influence of any mode of artificial stimulation (except heat), which is of a constant character and of an intensity sufficiently low not to produce tetanus. And as I have proved that the rhythm thus artificially produced is almost certainly due to the alternate process of exhaustion and recovery which I have explained, there can scarcely be any doubt that in the natural rhythm this process plays an important part, particularly as we find that temperature and gases exert the same influences on the one rhythm as they do on the other. Again, as an additional reason for recognizing the part which the contractile tissues probably play in the production of rhythm, I have pointed to the fact that in the great majority of cases in which rhythmic action occurs the presence of ganglia cannot be suspected. For it is among the lower forms of life, where ganglia are certainly absent, and where the functions of stimulation and contraction appear to be blended and diffused, that rhythmic action is of the most frequent occurrence; and it is obvious with how much greater difficulty the resistance theory is here beset than is the one I now propose. Granted a diffused power of stimulation with a diffused power of response, and I see no essential difference between the rhythmic motions of the simplest organism and those of a deganglionated Medusa in acidulated water. Lastly, the facts relating to the tonus of the manubrium in Sarsia furnish very striking, and I think almost conclusive, proof of the theory which I have advanced.

CHAPTER IX.

POISONS.

1. *Chloroform.*—My observations with regard to the distribution of nerves in Sarsia led me to investigate the order in which these connections are destroyed, or temporarily impaired, by anæsthetics. The results, I think, are worth recording. In Sarsia the following phases always mark the progress of anæsthesia by chloroform, etc.—1. Spontaneity ceases. 2. On now nipping a tentacle, pulling the manubrium, or irritating the bell, a *single* locomotor contraction is given in answer to every stimulation. (In the unanæsthesiated animal a *series* of such contractions would be the result of such stimulation.) 3. After locomotor contractions can no longer be elicited by stimuli, nipping a tentacle or the margin of the bell has the effect of causing the manubrium to contract. 4. After stimulation of any part of the nectocalyx (including tentacles) fails to produce response in any part of the organism, the manubrium will continue its response to stimuli applied directly to itself.

2. *Nitrite of Amyl.*—On Sarsia the effect of this

agent is much the same as that of chloroform—the description just given being quite as applicable to the effects of the nitrite as to those of chloroform. Before the loss of spontaneity supervenes, the rate of the rhythm is increased, while the strength of the pulsations is diminished.

Tiaropsis diademata, from the fact of its presenting a very regular rhythm and being but of small size, is a particularly suitable species upon which to conduct many experiments relating to the effect of poisons. On this species the nitrite in appropriate (*i.e.* in very small) doses first causes irregularity and enfeeblement of the contractions, together with quickening of the rhythm. After a short time, a gradual cessation of the swimming motions becomes apparent—these motions dying out more gradually, for example, than they do under the influence of chloroform. Eventually each pulsation is marked only by a slight contraction of the muscular tissue in the immediate neighbourhood of the margin. If the dose has been stronger, however, well-marked spasmodic contractions come on and obliterate such gradual working of the poison. In all cases irritability of all parts of the animal persists for a long time after entire cessation of spontaneous movements—perhaps for three or four minutes in not over-poisoned animals; but eventually it too disappears. On being now transferred to normal sea-water, the process of recovery is slower than it is after anæsthesiation by chloroform. It is interesting, moreover, to observe, that just as the power of co-ordination was the first thing to be affected by the nitrite, so it is the last thing to return during recovery.

3. *Caffein.*—The effects of caffein on Sarsia may be best studied by immersing the animals in a saturated sea-water solution of the substance. In such solutions the Medusæ float to the surface, in consequence of their lower specific gravity. I therefore used shallow vessels, in order that the margins of the nectocalyces might rest in the level of the water that was thoroughly saturated. The immediate effect of suddenly immersing Sarsia in such a solution is very greatly to increase the rate of the pulsations, and, at the same time, to diminish their potency. The appearance presented by the swimming motions is therefore that of a fluttering nature; and such motions are not nearly so effectual for progression as are the normal pulsations in unpoisoned water. This stage, however, only lasts for a few seconds, after which the spontaneous motions begin gradually to fade away. Soon they altogether cease, though occasionally one among a number of Sarsiæ confined in the same saturated solution will continue, even for several minutes after the first immersion, to give one or two very feeble contractions at long intervals. Eventually, however, all spontaneity ceases on the part of all the specimens, and now the latter will continue for a very long time to be sensitive to stimulation. At first *several* feeble locomotor contractions will be given in response to each stimulus; and as on the one hand these contractions never originate spontaneously, while, on the other hand, *paralyzed*

Sarsiæ never respond to a single stimulus with more than a single contraction, these multiple responses must, I think, be ascribed to a state of exalted reflex irritability. After a long exposure to the poison, however, only a single response is given to each stimulus; and still later all irritability ceases. On now transferring the Sarsiæ to unpoisoned water, recovery is effected even though the previous exposure has been of immensely long duration, *e.g.* an hour.

An interesting point with regard to caffein-poisoning of Sarsia is, that as soon as spontaneity ceases the tentacles and manubrium lose their tonus and become relaxed to their utmost extent. This is not the case with anæsthesiation by chloroform, even when pushed to the extent of suspending irritability. If, however, Sarsiæ which have been anæsthesiated to this extent in chloroform be suddenly transferred to a solution of caffein, the tentacles and manubrium may soon be seen to relax, and eventually these organs lose their tonus as completely as if the anæsthesia had from the first been produced by the caffein. Moreover in this experiment the irritability, which had been destroyed by the chloroform, returns in the solution of caffein—provided the latter be not quite saturated—though spontaneity of course remains suspended throughout.

The effects of graduating the doses of caffein may be stated in connection with another species, viz. Tiaropsis diademata. In a weak solution the effects are a quickening of the pulsations (*e.g.* from 64 to

120 per minute) together with a lessening of their force. On slightly increasing the dose, the pulsations become languid, and prolonged pauses supervene. If the dose is again somewhat strengthened, the pulsations become weaker and weaker, till they eventually cease altogether. The animal, however, is now in a condition of exalted reflex irritability; for its response to a single stimulus consists not merely, as in the unpoisoned animal, of a single spasm, but also, immediately after this, of a series of convulsive movements somewhat resembling swimming movements destitute of co-ordination. If the strength of the solution be now again increased, a stage of deeper anæsthesiation may be produced, in which the Medusa will only respond to each stimulation by a single spasm. In still stronger solutions, the only response is a single feeble contraction; while in a nearly saturated solution the animal does not respond at all. But even from a saturated solution Tiaropsis diademata will recover when transferred to unpoisoned water.

4. *Strychnia.*—The species of covered-eyed Medusa which I shall choose for describing the action of strychnia is Cyanæa capillata, which is most admirably adapted for experiments with this and some of the other alkaloid poisons, from the fact that in water kept at a constant temperature its pulsations are as regular as are those of a heart. After Cyanæa capillata has been allowed to soak for ten minutes or so in a weak sea-water solution of strychnia, unmistakable signs of irregularity in the

pulsations supervene. This irregularity then increases more and more, till at length it grows into well-marked convulsions. The convulsions manifest themselves in the form of extreme deviations from the rhythmical contractions so characteristic of Cyanæa capillata. Instead of the heart-like regularity with which systole and diastole follow one another in the unpoisoned animal, we now have periods of violent and prolonged systole resembling tonic spasm; and when the severity of this spasm is for a moment abated, it is generally renewed before the umbrella has had time again to become fully expanded. Moreover, the spasm itself is not of uniform intensity throughout the time it lasts; but while the umbrella is in a continuously contracted state, there are observable a perpetual succession of extremely irregular oscillations in the strength of the contractile influence. It is further a highly interesting fact that the convulsions are very plainly of a *paroxysmal* nature. After the umbrella has suffered a prolonged period of convulsive movements, it expands to its full dimensions, and in this form remains for some time in a state of absolute quiescence. Presently, however, another paroxysm supervenes, to be followed by another period of quiescence, and so on for hours. The periods of quiescence are usually shorter than are those of convulsion; for while the former seldom last more than forty seconds or so, the latter may continue uninterruptedly for five or six minutes. In short, Medusæ, when submitted to the influence of strychnia, exhibit all the symptoms of strychnia poisoning in the higher animals. Death, however, is always in the fully expanded form.

It seems desirable to supplement these remarks with a few additional ones on the effects of this poison on the naked-eyed Medusæ. In the case of Sarsia the symptoms of strychnia poisoning are not well marked, from the fact that in this species convulsions always take the form of locomotor contractions. The symptoms, however, are in some respects anomalous. They are as follows. First of all the swimming motions become considerably accelerated, periods of quiescence intervening between abnormally active bouts of swimming. By-and-by a state of continuous quiescence comes on, during which the animal is not responsive to tentacular irritation, but remains so to direct muscular irritation, giving one response to each direct stimulus. The tentacles and manubrium are much relaxed. In a sea-water solution just strong enough to taste bitter, this phase may continue for hours; in fact, till a certain opalescence of the contractile tissues—which it is a property of strychnia, as of most other reagents, to produce—has advanced so far as to place the tissues beyond recovery. If the exposure to such a solution has not been very prolonged, recovery of the animal in normal water is rapid. In a specimen exposed for two and a half hours to such a solution, recovery began in half an hour after restoration to normal water, but was never complete. In all cases, if the poisoning is allowed to pass beyond the stage at which response to direct muscular irritation ceases, the animal is dead.

On Tiaropsis indicans this poison has the effect of causing a general spasm, which would be undistinguishable from that which in this species results from general stimulation of any kind, were it not that there is a marked difference in one particular. For in the case of strychnia poisoning, the spasm, while it lasts, is not of uniform intensity over all parts of the nectocalyx; but now one part and now another part or parts are in a state of stronger contraction than other parts, so that, as a general consequence, the outline of the nectocalyx is continually changing its form. Moreover, in addition to these comparatively slow movements, there is a continual twitching observable throughout all parts of the nectocalyx. Each individual twitch only extends over a small area of the contractile tissue; but in their sum their effect is to throw the entire organ into a sort of shivering convulsion, which is superimposed on the general spasm. After a time the latter somewhat relaxes, leaving the former still in operation, which, moreover, now assumes a paroxysmal nature—the convulsions consisting of strong shudders and frequent spasms with occasional intervals of repose.

In the case of Tiaropsis diademata the action of strychnia is very similar, with the exception that there is no *continuous* spasm, although *occasional* ones occur amid the twitching convulsions. After a time, however, all convulsions cease, and the animal remains quiescent. While in this condition its reflex excitability is abnormally increased, as shown by the fact that even a gentle touch will bring on, not merely a single responsive spasm, as in the unpoisoned animal, but a whole series of successive spasms, which are often followed by a paroxysm of twitching convulsions. The condition of exalted reflex irritability is thus exceedingly well marked. Recovery in normal water at this stage is rapid, the motions being at first characterized by a want of co-ordination, which, however, soon passes off.

5. *Veratrium.*—In Sarsia the first effect of this poison is to increase the number and potency of the contractions; but its later effect is just the converse, there being then prolonged periods of quiescence, broken only by very short swimming bouts consisting of feeble contractions. The feebleness of the contractions gradually becomes more and more remarkable, until at last it is with great difficulty that they can be perceived at all; indeed, the progressive fading away of the contractions into absolute quiescence is so gradual that it is impossible to tell exactly when they cease. During the quiescent stage the animal is for the first time insensible both to tentacular and to direct stimulation of the contractile tissues. That the gradual dying out of the strength of the contractions is not altogether due to the progressive advance of central paralysis, would seem to be indicated by the fact that contractions in response to direct stimulation of the con'ractile tissues are no more powerful, at any given stage of the poisoning, than are either responses to tentacular stimulation or the spontaneous contractions. Still, as we shall immediately

sec, in the various species of Tiaropsis, irritability persists after cessation of the spontaneous contractions. In Sarsia the nervous connections between the tentacles and manubrium, and also between the tentacles themselves, are not impaired during the time that the bell is motionless; and even when the irritability of the bell has quite disappeared as regards any kind of stimulation, the manubrium and tentacles will continue responsive to stimuli applied either directly to themselves or to any part of the neuro-muscular sheet of the bell.

The convulsions due to the action of veratrium are well marked in the various species of the genus Tiaropsis. They consist of violent fluttering motions without any co-ordination; but there are no spasms, as in the case of strychnia poisoning. After the convulsions have lasted for some time, a quiescent stage comes on, during which the animal remains responsive to stimulation, though not abnormally so. Recovery in unpoisoned water is rapid, the movements being at first marked by an absence of co-ordination.

6. *Digitalin.*—The first effect of this poison on Sarsia is to quicken the swimming motions, and then to enfeeble them progressively till they degenerate into mere spasmodic twitches. The manubrium and tentacles are now strongly retracted, while the nectocalyx is drawn together so as to assume an elongated form. The latter is now no longer responsive either to tentacular or to direct stimulation; but the tentacles and manubrium both remain responsive to stimuli applied either directly to

themselves or to the neuro-muscular tissue of the bell. Death always takes place in very strong systole; and as this is an exceedingly unusual thing in the case of Sarsia, there can be no doubt that, in this respect, the action of the digitalin is different on the Medusæ from what it is on the heart.

On the various species of Tiaropsis, digitalin at first causes acceleration of the swimming movements, with great irregularity and want of co-ordination. Next, strong and persistent spasms supervene, which give the outline of the necto-calyx an irregular form; and every now and then this unnatural spasm gives place to convulsive swimming motions. Evidently, however, the spasm becomes quite persistent and excessively strong. The manubrium of Tiaropsis indicans crouches to its utmost, and the animal dies in strong systole.

7. *Atropin.*—In the case of Sarsia atropin causes convulsive swimming motions. The systoles next become feeble, and finally cease. The nectocalyx is now somewhat drawn together in persistent systole, with the manubrium and tentacles strongly retracted. Muscular irritability remains after tentacular irritability has disappeared, but it is then decidedly enfeebled.

In the various species of Tiaropsis the convulsions are strongly pronounced. They begin as mere accelerations of the natural swimming motions, but soon grow into well-marked convulsions, consisting of furious bouts of irregular systoles following one another with the utmost rapidity, and wholly without co-ordination. Occasionally these movements are interrupted by a violent spasm, on which strong shuddering contractions are superimposed.

8. *Nicotin.*—On dropping Sarsia into a sea-water solution of nicotin of appropriate strength, the animal immediately goes into a violent and continuous spasm, on which a number of rapidly succeeding minute contractions are superimposed. The latter, however, rapidly die away, leaving the nectocalyx still in strong and continuous systole; tentacles and manubrium are retracted to the utmost. Shortly after cessation of spontaneity, the bell is no longer responsive to tentacular stimulation, but remains for a considerable time responsive to direct stimulation of its own substance; eventually, however, all irritability disappears, while the tentacles and manubrium relax. On transferring the animal to normal water, muscular irritability first returns, and then central, as shown by the earlier response of the bell to direct than to tentacular stimulation; but if the animal has been poisoned heavily enough to have had its muscular irritability suspended, it is a long time before central irritability returns. Soon after central irritability has returned, the animal begins to show feeble signs of spontaneity, the motions being exceedingly weak, with long intervals of repose; but the degree of such feebleness depends on the length of time during which the animal has previously been exposed to the poison; thus in a specimen which had been removed from the poison immediately after the disappearance of reflex

irritability had supervened, recovery began in ten minutes after re-immersion, and was complete in half an hour.

In Tiaropsis the symptoms of nicotin poisoning are also well marked. When gradually administered, the first effect of the narcotic is a complete loss of co-ordination in the swimming motions. A slight increase of the dose brings about a tonic spasm, which differs from the natural spasm of these animals—(*a*) in being stronger, so that the nectocalyx becomes bell-shaped rather than square, (*b*) in being much more persistent, and (*c*) in undergoing variations in its intensity from time to time, instead of being a contraction of uniform strength; thus the spasm temporarily affects some parts of the nectocalyx more powerfully than other parts, so that the organ may assume all sorts of shapes. Such distortions proceed even further under the influence of nicotin than under that of strychnine, etc. Sometimes, for instance, one quadrant will project in the form of a pointed promontory; at other times two adjacent or opposite quadrants will thus project, and occasionally all four will do so, the animal thus becoming star-shaped. Sometimes, again, one quadrant will be less contracted than the other three, while at other times more or less slight relaxations affect numerous parts of the bell, its margin being thus rendered sinuous, though more or less violently contracted in all its parts. This state of violent spasm lasts for several minutes, when it gradually passes off, the nectocalyx relaxing into the form of a deep bowl and remaining quite

passive, except that every now and then one part
or another of the margin is suddenly contracted in
a semilunar form. By-and-by, however, even these
occasional twitches cease, and the animal is now
insensible to all kinds of stimulation. Recovery in
normal water is gradual, and marked in its first
stage by the occasional retractions of the margin
last mentioned. At about this stage also, or sometimes slightly later, the animal first becomes responsive to stimulation; and it is interesting to note
that the response is performed, not by giving a
general spasm as would the unpoisoned animal, but
by folding in the part irritated—an action which
very much resembles, on the one hand, the spontaneous convulsive movements just described, and,
on the other, the response which is given to stimulation by the unpoisoned bell when gently irritated
after removal of its margin. After these stages
there supervenes a prolonged period of quiescence,
during which the animal remains normally responsive to stimulation. Spontaneity may not return
for several hours, and, after it does return, the
animal is in most cases permanently enfeebled.
Indeed, on all the species of Medusæ, nicotin, both
during its action and in its subsequent effects, is
the most deadly of all the poisons I have tried.

9. *Morphia.*—The anæsthesiating effects of morphia are as decided as are those of chloroform. I shall
confine myself to describing the process of anæsthesiation in the case of Aurelia aurita in an extract
from my notes. "A very vigorous specimen, having
twelve lithocysts, was placed in a strong sea-water

solution of morphia. Half a minute after being introduced commencement of torpidity ensued, shown by contractions becoming fewer and feebler In one minute the feeble impulses emanating from the prepotent lithocyst failed to spread far through the contractile tissue, appearing to encounter a growing resistance. Eventually this resistance became so great that only a very small portion of contractile tissue in the immediate neighbourhood of the lithocyst contracted, and this in a very slow and feeble way. Two minutes after immersion even these partial contractions entirely ceased, and soon afterwards all parts of the animal were completely dead to stimulation. Recovery in normal water slower than that after chloroform, but still soon quite complete. Repeated experiment on this individual four times without injury."

10. *Alcohol.*—The solution must be strong to cause complete intoxication. The first effect on Sarsia is to cause a great increase in the rapidity of the swimming motions—so much so, indeed, that the bell has no time to expand properly between the occurrence of the successive systoles, which, in consequence, are rendered feeble. These motions gradually die out, leaving the animal quite motionless. The nectocalyx is now responsive to stimuli applied at the tentacles, and sometimes two or three contractions will follow such a stimulus, as if the spontaneity of the animal were slightly aroused by the irritation. Soon, however, only one contraction is given in response to every tentacular irritation, and by-and-by this also ceases—the Medusa

being thus no longer responsive to central stimulation. It remains, however, for a long time responsive to stimulation of the neuro-muscular sheet; indeed, the strength of the alcohol solution must be very considerable before loss of muscular irritability supervenes. It may thus be made to do so, however; and on then transferring the animal to normal water, recovery begins in from three minutes to a quarter of an hour. The first contractions are very feeble, with long intervals of repose; but gradually the animal returns to its normal state.

The above remarks apply also to Tiaropsis. In Tiaropsis indicans the manubrium recovers in normal water sooner than the nectocalyx. Both in Sarsia and Tiaropsis the manubrium and tentacles are retracted while exposed to alcohol, and, after transference to normal sea-water, the animals float on the surface, presumably in consequence of their having imbibed some of the spirit. The period during which flotation lasts depends, (a) on the strength of the alcohol solution used, and (b) on the time of exposure to its influence. It may last for an hour or more; but in no case is recovery complete till some time after the flotation ceases.

11. *Curare.*—Curare had already been tried upon Medusæ, and was stated to have produced no effects; it is therefore especially desirable that I should first of all describe the method of exhibiting it which I employed.

Having placed the medusa to be examined in a flat-shaped beaker, I filled the latter to overflowing with sea-water. I next placed the beaker in a large basin, into which I then poured sea-water until the level was the same inside and outside the beaker, *i.e.* until the two bodies of water all but met over the brim. Having divided the medusa across its whole diameter, with the exception of a small piece of marginal tissue at one side to act as a connecting link between the two resulting halves, I transferred one of these halves to the water in the basin, leaving the other half still in the beaker— the marginal tissue which served to unite the two halves being thus supported by the rim of the beaker. Over the minute portion of the marginal tissue which was thus of necessity exposed to the air, I placed a piece of blotting-paper which dipped freely into the sea-water. Lastly, I poisoned the water in the beaker with successive doses of curare solution.

The results obtained by this method were most marked and beautiful. Previous to the administration of the poison both halves of the medusa were of course contracting vigorously, waves or contractile influence now running from the half in the beaker to the half in the basin, and now *vice versâ*. But after the half in the beaker had become effectually poisoned by the curare, all motion in it completely ceased, the other, or unpoisoned half, continuing to contract independently. I now stimulated the poisoned half by nipping a portion of its margin with the forceps. Nothing could be more decided than the result. It will be remembered that when any part of Staurophora laciniata is pinched with the forceps or otherwise irritated, the motion of the whole body which ensues is totally different from that of an ordinary locomotor contraction—all parts folding together in one very strong and long-protracted systole, after which the diastole is very much slower than usual. Well, on nipping any portion of the poisoned half of Staurophora laciniata, this half remained absolutely motionless, while the unpoisoned half, though far away from the seat of irritation, immediately ceased its normal contractions, and folded itself together in the very peculiar and distinctive manner just described. This observation was repeated a number of times, and, when once the requisite strength of the curare solution had been obtained, always with the same result. The most suitable strength I found to be 1 in 2500, in which solution the poisoned half required to soak for half an hour.

I also tried the effect of this poison on the covered-eyed Medusæ, and have fairly well satisfied myself that its peculiar influence is likewise observable in the case of this group, although not in nearly so well-marked a manner.

It has further to be stated that when the poisoned half is again restored to normal sea-water, the effects of curare pass off with the same rapidity as is observable in the case of the other poisons which I have tried. Thus, although an exposure of half an hour to the influence of curare of the strength named is requisite to destroy the motor power in the case of Staurophora laciniata, half a minute is sufficient to ensure its incipient return when the animal is again immersed in unpoisoned water.

It is also to be observed that a very slight degree of *over*-poisoning paralyzes the transmitting system as well as the responding one; so that if any one should repeat my observation, I must warn him against drawing erroneous conclusions from this fact. Let him use weak solutions with prolonged soaking, and by watching when the voluntary motions in the poisoned half first cease, he need experience no difficulty in obtaining results as decided as it is possible for him to desire.

12. *Cyanide of Potassium.*—On Sarsia the first effect is to quicken the contractions and then to enfeeble them. The animal assumes an elongated form, as already described under atropin. Spontaneity ceases very rapidly even in weak solutions; and for an exceedingly short time after it has done so, the bell continues responsive both to tentacular and to direct stimulation. For a long time after the bell ceases to respond to any kind of stimulation, the nervous connections between the tentacles and between the tentacles and manubrium remain intact, as also do the nervous connections of these organs with all parts of the bell. This interesting fact is rendered apparent, first, by stimulating a tentacle and observing that all the four tentacles and the manubrium respond; and, second, by irritating any part of the neuro-muscular sheet of the bell, and observing that while the latter does not respond both the tentacles and the manubrium retract. Recovery from this stage occupies several hours.

In the case of Tiaropsis the convulsions are, as usual, more pronounced, being marked by the occurrence of a gradually increasing spasm, which differs from a normal spasm in the respects already described under strychnia. In all the species both of Sarsia and Tiaropsis, the manubrium and tentacles are retracted during exposure to this poison.

Remarks.

The above comprises all the poisons which I have tried, and I think that all the observations taken together show a wonderful degree of resemblance between the actions of the various poisons on the Medusæ and on the higher animals—a general fact which is of interest, when we remember that in these nerve-poisons we possess, as it were, so many tests wherewith to ascertain whether nerve-tissue, where it first appears upon the scene of life, presents the same fundamental properties as it does in the higher animals. And these observations show that such is the case. When the physiologist bears in mind that in Sarsia we have the means of testing the comparative influence of any poison on the central, peripheral, and muscular systems respectively,* he will not fail to appreciate the significance of these observations. In reading over the whole list he will meet with an anomaly here and

* The method of comparison consists, as will already have been gathered from the perusal of the foregoing sections, in:—first, stimulating the tentacles, and observing whether this is followed by such a discharge of the attached ganglion as causes the bell to contract; next, stimulating the bell itself, to ascertain whether the muscular irritability is impaired; and, lastly, stimulating either the tentacles or the bell, to observe whether the reciprocal connections between tentacles, bell, and manubrium are uninjured.

there; but, on the whole, I do not think he can fail to be satisfied with the wonderfully close adherence which is shown by these elementary nervous tissues to the rules of toxicology that are followed by nervous tissues in general. In one respect, indeed, there is a conspicuous and uniform deviation from these rules; for we have seen that in the case of every poison mentioned more or less complete recovery takes place when the influence of the poison has been removed, even though this has acted to the extent of totally suspending irritability. In other words, there is no poison in the above list which has the property, when applied to the Medusæ, of destroying life till long after it has destroyed all signs of irritability. What the cause of this uniform peculiarity may be is, of course, conjectural; but I may suggest two considerations which seem to me in some measure to mitigate the anomaly. In the first place, we must remember that in the Medusæ there are no nervous centres of such vital importance to the organism that any temporary suspension of their functions is followed by immediate death. Therefore, in these animals, the various central nerve-poisons are at liberty, so to speak, to exert their full influence on all the excitable tissues without having the course of their action interrupted by premature death of the organism, which in higher animals necessarily follows the early attack of the poison on a vital nerve-centre. Again, in the second place, we must remember that the method of administering the above-mentioned poisons to the Medusæ was very different from that which we employ when administering them to other animals; for, in the case of the Medusæ, the neuro-muscular tissue is spread out in the form of an exceedingly tenuous sheet, so that when the animal is soaking in the poisoned water every portion of the excitable tissue is equally exposed to its influence; and that the action of a poison is greatly modified by such a difference in the mode of its administration has been proved by Professor Gamgee, who found that when a frog's muscle is allowed to soak in a solution of vanadium, etc., it loses its irritability, while this is not the case if the poison is administered by means of the circulation.

I may further observe that in the case of all poisons I have tried, the time required for recovery after the animal is restored to normal water varies immensely. The variations are chiefly determined by the length of time during which the animal has been exposed to the influence of the poison, but also, in a lesser degree, by the strength of the solution employed. To take, for instance, the case of caffein or chloroform, if Sarsiæ are transferred to normal water after they first cease to move, a few seconds are enough to restore their spontaneity; whereas, if they are allowed to remain in the poisoned water for an hour, they may not move for one or two hours after their restoration to unpoisoned water. In consequence of such great variations occurring from these causes, I was not able to compare the action of one poison with that of another in respect of the time required for effects of poisoning to pass away.

I shall conclude all I have to say upon the subject of poisons by stating the interesting fact, that if any of the narcotic or anæsthesiating agents be administered to any portion of a contractile strip cut from the umbrella of Aurelia aurita in the way already described, the rate of the contraction-waves is first progressively slowed, and eventually their passage is completely blocked at the line where the poisoned water begins. Upon now restoring the poisoned portion of the contractile strip to normal sea-water the blocking is gradually overcome, and eventually every trace of it disappears.*

The contractile wave may be blocked by poisons in another way. A glance at Fig. 11 will show that a circumferential strip cut from the umbrella of Aurelia aurita is pervaded transversely by a number of nutrient tubes, which have all been cut through by the section. At the side of the strip, therefore, furthest from the margin there are situated a number of open ends of these nutrient tubes. Now, on injecting any of the narcotic poisons into

* In conducting this experiment, care must be taken not to exert the slightest pressure on any part of the strip. The method I adopted, therefore, was to have a vessel with a very deep furrow on each of its opposite lips. Upon filling this vessel to the level of these furrows with the poisoned water, and then immersing the whole vessel in ordinary sea-water up to the level of its brim, some of the poisoned water of course passed through the open furrows. The external body of water (*i.e.* the normal sea-water containing the animal) was therefore made proportionally very large, so that the slight escape of poison into it did not affect the experiment. On now passing the portion of the strip to be poisoned through the opposite furrows, it was allowed to soak in the poison while freely floating, and so without suffering pressure in any of its parts.

one of these open ends, the fluid of course permeates the whole tube, and the contraction-wave becomes blocked at the transverse line occupied by the tube as effectually as if the contractile strip had been cut through at that line.

A glance at Fig 10, again, will show that each lithocyst is surrounded by one of these nutrient tubes. Upon injecting this tube, therefore, in a contractile strip, the effect of the poison may be exerted on the lithocyst more specially than it could be by any other method of administration. In view of recent observations concerning the effects of curare on the central nervous masses of higher animals, it may be worth while to state that a discharging lithocyst of Aurelia aurita, when thus injected with curare, speedily ceases its discharges. This fact alone, however, would not warrant any very trustworthy conclusions as to the influence of curare upon discharging centres; for it is not improbable that the paralyzing effects may here be due to the influence of the poison on the surrounding contractile tissue.

It is interesting to observe that if the discharging lithocyst be injected with chloroform, or a not too strong solution of morphia, it recovers in the course of a night. With alcohol the first effects of the injection are considerably to accelerate the frequency and to augment the potency of the discharges; but the subsequent effects are a gradual diminution in the frequency and the vigour of these discharges, until eventually total quiescence supervenes. In the course of a few hours, however, the torpidity wears away, and finally the medusa returns to its normal state.*

* Since the above results on the effects of poisons were published in my Royal Society papers, Dr. Krukenberg has conducted a research upon "comparative toxicology," in which he has devoted the larger share of his attention to the Medusæ. While expressing my gratification that when he adopted my methods he succeeded in confirming my results, I may observe that the criticism which he somewhat bluntly passes upon the latter is not merely unwarranted, but based upon a strange misconception of a well-known principle in the physiology of nerves and muscles. This criticism is that these results as published by me are worthless and "a dead chapter in science," because I failed to prove that it was the nervous (as distinguished from the muscular) elements which were effected by the various poisons. In his opinion this distinction can only be made good by employing electrical stimulation upon the sub-umbrella tissue when this has lost its spontaneity under the influence of poisons : if a response ensues which does not ensue when the tissue is stimulated mechanically, he regards the fact as proof that the muscular tissue remains unaffected while the nervous tissue has been rendered functionless.

Now, in the first place, I have here to show that there is, as I have said, a fundamental error touching a well-known principle of physiology. So far as there is any difference between the excitability of nerve and muscle with respect to mechanical and electrical stimulation, it is the precise converse of that which Dr. Krukenberg supposes; it is not *muscle*, but *nerve*, which is the more sensitive to electrical stimulation—by which I understand him to mean the induction shock. The remarkable transposition of Dr. Krukenberg's ideas upon this matter does not affect the results of his observations upon the action of the various poisons; it only renders fatuous his criticism of these same results as previously published by me.

In the next place, I have to observe that in all my experiments I tried, as he subsequently tried, both kinds of stimulation, and also the constant current; but I soon found that even when one went to work with one's ideas upon the subject in a non-inverted position, no trustworthy inference could be drawn in favour of the muscular elements alone remaining uninjured from the bare

Physiological Effects of Fresh Water on the Medusæ

As fresh water exerts a very deadly influence on the Medusæ, this seems the most appropriate place for describing its action. Such a description has already been given by Professor L. Agassiz, but it is erroneous. He writes, "Taking up in a spoonful of sea-water a fresh Sarsia in full activity, when swimming most energetically, and emptying it into a tumbler full of fresh water of the same temperature, the little animal will at once drop like a ball

fact that after the poisoning the neuro-muscular tissue often behaved differently towards different kinds of stimulation.

Further, in the particular case of my experiments with curare—against which Dr. Krukenberg's remarks are chiefly directed on the ground that I did not prove the paralysis to be a merely muscular effect—I succeeded in obtaining very much better proof of the poison acting on the nervous elements, to the exclusion of the muscular, than I could have obtained by any process of inference, however good; that is to say, I obtained direct proof. It appears to me that Dr. Krukenberg must have failed to understand the English of the following sentences: "On nipping any portion of the poisoned half of Staurophora laciniata, this half remained *absolutely motionless*, while the unpoisoned half, *though far away from the seat of irritation*, immediately ceased its normal contractions, and folded itself together in the very peculiar and distinctive manner just described," *i.e.* "in one very strong and long-protracted systole." For the rest, see note on page 232.

Lastly, while again expressing my satisfaction that on all matters of fact our results are in full harmony, I may be allowed to remark that in my opinion his deductions, as embodied in his schema of the inferred innervation of Medusæ, are very far in advance of anything that is justified by observation. (See, for this elaborate schema, in which there are represented volitional, motor, reflex, and inhibitory centres, as well as a clearly defined system of sensory and motor nerves, "Vergleichend-Physiologische Studien, dritte Abtheilung," p. 141: Heidelberg, 1880.)

to the bottom of the glass and remain for ever motionless—killed instantaneously by the mere difference of the density of the two media."* As regards the appearance presented by Sarsia when subjected to "this little experiment," the account just quoted is partly correct; but Professor Agassiz must have been over-hasty in concluding that, because the animals seemed to be thus "killed instantaneously," such was really the case. Nothing, indeed, could be more natural than his conclusion; for not only is the contrast between the active swimming motions of the Sarsia in the sea-water and their cessation of all motion in the fresh water very suggestive of instantaneous death; but a short time after immersion in the latter their contractile tissues, as Professor Agassiz observed, become opalescent and whitish. Nevertheless, if he had taken the precaution of again transferring the Sarsia to sea-water, he would have found that the previous exposure to fresh water had not had the effect which he ascribes to it. After a variable time his specimens would have resumed their swimming motions; and although these might have had their vigour somewhat impaired, the animals would have continued to live for an indefinite time—in fact, quite as long as other specimens which had never been removed from the sea-water. Even after five minutes' immersion in fresh water, Sarsiæ will revive feebly on being again restored to sea-water, although it may be two or three hours before they do so; they may then, how-

* "Mem. American Acad. Arts and Sciences," 1850, p. 229.

ever, live as long as other specimens. In many cases Sarsiæ will revive even after ten minutes' exposure; but the time required for recovery is then very long, and the subsequent pulsations are of an exceedingly feeble character. I never knew a specimen survive an exposure of fifteen minutes.* In not a few cases, after immersion in fresh water, the animal continues to pulsate feebly for some little time; and, in all cases, irritability of the contractile tissues persists for a little while after spontaneity has ceased. The opalescence above referred to principally affects the manubrium, tentacles, and margin of the nectocalyx. While in fresh water the manubrium and tentacles of Sarsia are strongly retracted.

Thinking it a curious circumstance that the mere absence of the few mineral substances which occur in sea-water should exert so profound and deadly an influence on the neuro-muscular tissues of the Medusæ, I was led to try some further experiments to ascertain whether it is, as Agassiz affirms, to the mere difference in density between the fresh and the sea water, or to the absence of the particular mineral substances in question, that the deleterious influence of fresh water is to be ascribed. Although

---

* The covered-eyed Medusæ survive a longer immersion than the naked-eyed—Aurelia aurita, for instance, requiring from a quarter to half an hour's exposure before being placed beyond recovery. Moreover, the cessation of spontaneity on the first immersion is not so sudden as it is in the case of the naked-eyed Medusæ—the pulsations continuing for about five minutes, during which time they become weaker and weaker in so gradual a manner that it is hard to tell exactly when they first cease.

my experiments lead to no very instructive conclusion, they are, I think, worth stating.

I first tried dissolving chloride of sodium in fresh water till the latter was of the same density as sea-water. Sarsiæ dropped into such a solution continued to live for a great number of hours; but they were conspicuously enfeebled, keeping for the most part at the bottom of the vessel, and having the vigour of their swimming motions greatly impaired. The tentacles and manubrium were strongly retracted, as in the case of exposure to fresh water, and the tissues also became slightly opalescent. Thinking that perhaps a fairer test would be only to add as much chloride of sodium to the fresh water as occurs in sea-water, I did so; but the results were much the same. On now adding sulphate of magnesium, however, to the amount normally present in sea-water, the Sarsiæ became more active. I next tried the effects of chloride of sodium dissolved in fresh water to the point of saturation, or nearly so. The Sarsiæ, of course, floated to the surface, and they immediately began to show symptoms of torpidity. The latter became rapidly more and more pronounced, till spontaneity was quite suspended. The animals, however, were not dead, nor did they die for many hours, their irritability continuing unimpaired, although their spontaneity had so completely ceased. The tentacles and manubrium were exceedingly relaxed, which is an interesting fact, as being the converse of that which occurs in water containing too small a proportion of salt. Lastly, to give the density hypothesis a still more complete trial, I dissolved various neutral salts and other substances, such as sugar, etc., in fresh water till it was of the density of sea-water; but in all cases, on immersing Sarsiæ in such solutions, death was as rapid as that which followed their immersion in fresh water.

The Fresh-water Medusa.

On June 10, 1880, it was noticed that the fresh water in the large tank of the lily-house of the Royal Botanical Society, Regent's Park, was swarming with a small and active species of Medusa, previously unknown to science—it being, indeed, at that time unknown to science that any species of Medusa inhabited fresh water, although it was well known that some of the other Hydrozoa do so. Examination showed that the new species belonged to the order Trachomedusæ, and the Petasidæ of Haeckel's classification—its nearest known relative, according to Professor Ray Lankester, being the genus Aglauropsis, which occurs on the coast of Brazil. The Medusa was called Limnocodium (λίμνη, a pond, and κώδων, a bell) sorbii by Professors Allman and Lankester. I am indebted to the kindness of Professor Allman for permission to reproduce his drawing of the animal. Fig. 31.) It is remarkable that, although this Medusa has reappeared every June in the same tank, no one has yet succeeded in tracing its life-history. Nor is it known from what source the tank first became impregnated with this organism. No doubt the germs must

POISONS. 243

have been conveyed by the roots or leaves of some tropical plant that at some time was placed in the

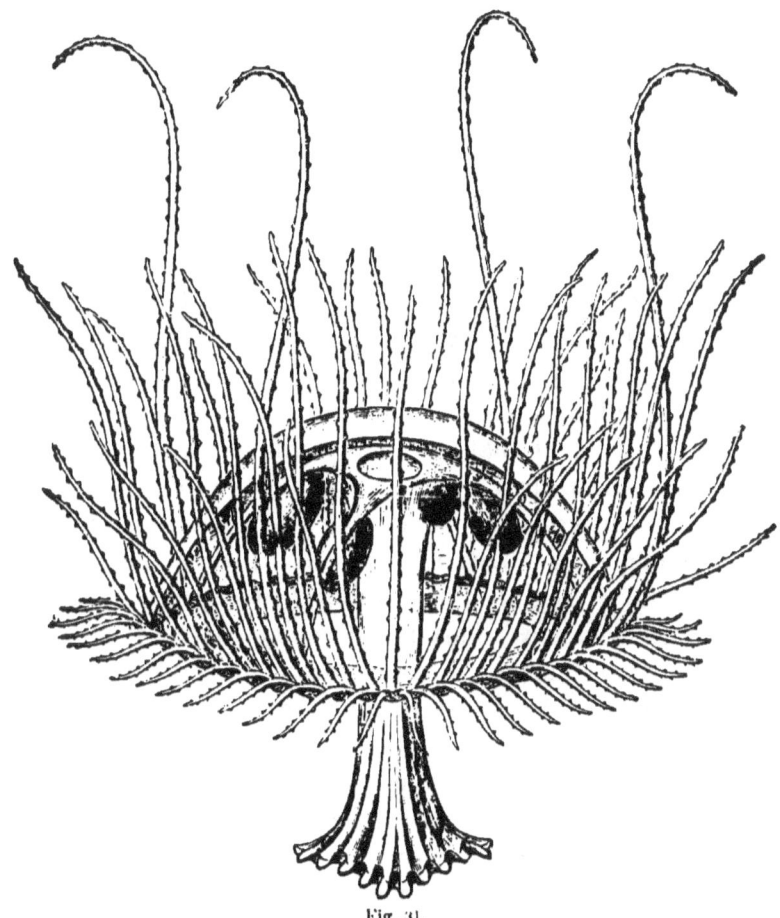

Fig. 31.

tank; but the Botanical Society has no record of any plant which can be pointed to as thus having probably served to import the organism.

I shall now proceed to give an account of my

observations on the physiology of this interesting animal, by quoting *in extenso* my original paper upon the subject (*Nature*, June 24, 1880). Before doing so, however, I may state that Professors A. Agassiz, Moseley, and others have since informed us that sundry species of sea-water Medusæ have been observed by them living and thriving in the brackish waters of estuaries—a fact which strongly corroborates the inference at the end of the present paper.

"The natural movements of the Medusa precisely resemble those of its marine congeners. More particularly, these movements resemble those of the marine species which do not swim continuously, but indulge in frequent pauses. In water at the temperature of that in the Victoria lily-house (85° Fahr.), the pauses are frequent, and the rate of the rhythm irregular, suddenly quickening and suddenly slowing even during the same bout, which has the effect of giving an almost intelligent appearance to the movements. This is especially the case with young specimens. In colder water (65° to 75°) the movements are more regular and sustained; so that, guided by the analogy furnished by my experiments on the marine forms, I infer that the temperature of the natural habitat of this Medusa cannot be so high as that of the water in the Victoria lily-house. In water of that temperature the rate of the rhythm is enormously high, sometimes rising to three pulsations per second. But by progressively cooling the water, this rate may be progressively lowered, just as in the case of the marine species; and in water at 65°, the maximum rate that I have observed is eighty pulsations per minute. As the temperature at which the greatest activity is displayed by the fresh-water species is a temperature so high as to be fatal to all the marine species which I have observed, the effects of cooling are, of course, only parallel in the two cases when the effects of a series of higher temperatures in the one case are compared with those of a series of lower temperatures in the other. Similarly, while a temperature of 70° is fatal to all the species of marine Medusæ which I have examined, it is only a temperature of 100° that is fatal to the fresh-water species. Lastly, while the marine species will endure any degree of cold without loss of life, such is not the case with the fresh-water species. Marine Medusæ, after having been frozen solid, will, when gradually thawed out, again resume their swimming movements; but this fresh-water Medusa is completely destroyed by freezing. Upon being thawed out, the animal is seen to have shrunk into a tiny ball, and it never again recovers either its life or its shape.

"The animal seeks the sunlight. If one end of the tank is shaded, all the Medusæ congregate at the end which remains unshaded. Moreover, during the daytime they swim about at the surface of the water; but when the sun goes down they subside, and can no longer be seen. In all these habits they resemble many of the sea-water species. They are themselves non-luminous.

"I have tried on about a dozen specimens the effect of excising the margin of the nectocalyx. In the case of all the specimens thus operated upon, the result was the same, and corresponded precisely with that which I have obtained in the case of marine species; that is to say, the operation produces immediate, total, and permanent paralysis of the nectocalyx, while the severed margin continues to pulsate for two or three days. The excitability of a nectocalyx thus mutilated persists for a day or two, and then gradually dies out, thus also resembling the case of the marine naked-eyed Medusæ. More particularly, the excitability resembles that of those marine species which sometimes respond to a single stimulation with two or three successive contractions.

"A point of specially physiological interest may be here noticed. In its unmutilated state the fresh-water Medusa exhibits the power of localizing with its manubrium a seat of stimulation situated in the bell; that is to say, when a part of the bell is nipped with the forceps, or otherwise irritated, the free end of the manubrium is moved over and applied to the part irritated. So far the movement of localization is precisely similar to that which I have previously described as occurring in Tiaropsis indicans (*Phil. Trans.*, vol. clxvii.). But further than this, I find a curious difference. For while in Tiaropsis indicans these movements of localization continue unimpaired after the margin of the bell has been removed, and will be ineffectually attempted even after the bell is almost entirely cut away from its connections with the manubrium, in the fresh-water Medusa these movements of localization cease after the extreme margin of the bell has been removed. For some reason or another the integrity of the margin here seems to be necessary for exciting the manubrium to perform its movements of localization. It is clear that this reason must either be that the margin contains the nerve-centres which preside over these localizing movements of the manubrium, or, much more probably, that it contains some peripheral nervous structures which are alone capable of transmitting to the manubrium a stimulus adequate to evoke the movements of localization. In its unmutilated state this Medusa is at intervals perpetually applying the extremity of its manubrium to one part or another of the margin of the bell, the part of the margin touched always bending in to meet the approaching extremity of the manubrium. In some cases it can be seen that the object of this co-ordinated movement is to allow the extremity of the manubrium— *i.e.* the mouth of the animal—to pick off a small particle of food that has become entangled in the marginal tentacles. It is therefore not improbable that in *all* cases this is the object of such movements, although in most cases the particle which is caught by the tentacles is too small to be seen with the naked eye. As it is thus no doubt a matter of great importance in the economy of the Medusa that its marginal tentacles should be very sensitive to contact with minute particles, so that a very slight stimulus applied to them should start the co-ordinated movements of localization, it is not surprising that the tentacular rim should present nerve-endings so far sensitive that only by their excitation can the reflex mechanism be thrown into action. But if such is the explanation in this case, it is curious that in Tiaropsis indicans every part of the bell should be equally capable of yielding a stimulus to a precisely similar reflex action.

"In pursuance of this point, I tried the experiment of cutting off *portions* of the margin, and stimulating the bell *above the portions of the margin which I had removed*. I found that in this case the manubrium did not remain passive as it did when the *whole* margin of the bell was removed; but that it made ineffectual efforts to find the offending body, and in doing so always touched some part of the margin which was still unmutilated. I can only explain this fact by supposing that the stimulus supplied to the mutilated part is spread over the bell, and falsely referred by the manubrium to some part of the sensitive—i.e. unmutilated—margin.

"But to complete this account of the localizing movements, it is necessary to state one additional fact which, for the sake of clearness, I have hitherto omitted. If any one of the four radial tubes is irritated, the manubrium will correctly localize the seat of irritation, whether or not the margin of the bell has been previously removed. This greater ease, so to speak, of localizing stimuli in the course of the radial tubes than anywhere else in the nectocalyx, except the margin, corresponds with what I found to be the case in Tiaropsis indicans, and probably has a direct reference to the distribution of the principal nerve-tracts.

"On the whole, therefore, contrasting this case of localization with the closely parallel case presented by Tiaropsis indicans, I should say that the two chiefly differ in the fresh-water Medusa, even when unmutilated, not being able to localize so promptly or so certainly, and in the localization being only performed with reference to the margin and radial tubes, instead of with reference to the whole excitable surface of the animal.

"All marine Medusæ are very intolerant of fresh water, and, therefore, as the fresh-water species must presumably have had marine ancestors,* it seemed an interesting question to determine how far this species would prove tolerant of sea-water. For the sake of comparison, I shall first briefly describe the effects of fresh water upon the marine species.† If a naked-eyed Medusa which is swimming actively in sea-water is suddenly transferred to fresh water, it will instantaneously collapse, become motionless, and sink to the bottom of the containing vessel. There it will remain motionless until it dies; but if it be again transferred to sea-water it will recover, provided that its exposure to the fresh water has not been of too long duration. I have never known a naked-eyed Medusa survive an exposure of fifteen minutes; but they may survive

---

* Looking to the enormous number of marine species of Medusæ, it is much more probable that the fresh-water species was derived from them than that they were derived from a fresh-water ancestry.

† For full account, see *Phil. Trans.*, vol. clxvii. pp. 744, 745.

an exposure of ten, and generally survive an exposure of five. But although they thus continue to live for an indefinite time, their vigour is conspicuously and permanently impaired; while in the fresh water irritability persists for a short time after spontaneity has ceased, and the tentacles and manubrium are strongly retracted.

"Turning now to the case of the fresh-water species, when first it is dropped into sea-water at 85° there is no change in its movements for about fifteen seconds, although the tentacles may be retracted. But then, or a few seconds later, there generally occurs a series of two or three tonic spasms, separated from one another by an interval of a few seconds. During the next half-minute the ordinary contractions become progressively weaker, until they fade away into mere twitching convulsions, which affect different parts of the bell irregularly. After about a minute from the time of the first immersion all movement ceases, the bell remaining passive in partial systole. There is now no vestige of irritability. If transferred to fresh water after five minutes' exposure, there immediately supervenes a strong and persistent tonic spasm, resembling rigor mortis, and the animal remains motionless for about twenty minutes. Slight twitching contractions then begin to display themselves, which, however, do not affect the whole bell, but occur partially. The tonic spasm continues progressively to increase in severity, and gives the outline of the margin a very irregular form; the twitching contractions become weaker and less frequent, till at last they altogether die away. Irritability, however, still continues for a time—a nip with the forceps being followed by a bout of rhythmical contractions. Death occurs in several hours in strong and irregular systole.

"If the exposure to sea-water has only lasted two minutes, a similar series of phenomena is presented, except that the spontaneous twitching movements supervene in much less time than twenty minutes. But an exposure of one minute may determine a fatal result a few hours after the Medusa has been restored to fresh water.

"Contact with sea-water causes an opalescence and eventual disintegration of the tissues, which precisely resemble the effects of fresh water upon the marine Medusæ. When immersed in sea-water this Medusa floats upon the surface, owing to its smaller specific gravity.

"In diluted sea-water (fifty per cent.) the preliminary tonic spasms do not occur, but all the other phases are the same, though extended through a longer period. In sea-water still more diluted (1 in 4 or 6) there is a gradual loss of spontaneity, till all movement ceases, shortly after which irritability also disappears; manubrium and tentacles expanded. After an hour's continued exposure, intense rigor mortis slowly and progressively developes itself, so that at last the bell has shrivelled almost to nothing. An exposure of a few minutes to this strength places the animal past recovery when restored to fresh water. In still weaker mixtures (1 in 8, or 1 in 10) spontaneity persists for a long time; but the animal gradually becomes less and less energetic, till at last it will only move in a bout of feeble pulsations when irritated. In still weaker solutions (1 in 12, or 1 in 15) spontaneity continues for hours, and in solutions of from 1 in 15, or 1 in 18, the Medusa will swim about for days.

"It will be seen from this account that the fresh-water Medusa is even more intolerant of sea-water than are the marine species of fresh water. Moreover, the fresh-water Medusa is beyond all comparison more intolerant of sea-water than are the marine species of brine; for I have previously found that the marine species will survive many hours' immersion in a saturated solution of salt. While in such a solution they are motionless, with manubrium and tentacles relaxed, so resembling the fresh-water Medusa shortly after being immersed in a mixture of one part sea-water to five of fresh; but there is the great difference that, while this small amount of salt is very quickly fatal to the fresh-water species, the large addition of salt exerts no permanently deleterious influence on the marine species.

"We have thus altogether a curious set of cross relations. It would appear that a much less profound physiological change would be required to transmute a sea-water jelly-fish into a jelly-fish adapted to inhabit brine, than would be required to enable it to inhabit fresh water. Yet the latter is the direction in which the modification has taken place, and taken place so completely that the sea-water is now more poisonous to the modified species than is fresh water to the unmodified. There can be no doubt that the modification was gradual—probably brought about by the ancestors of the fresh-water Medusa penetrating higher and higher through the brackish waters of estuaries into the fresh water of rivers—and it would, I think, be hard to point to a more remarkable case of profound physiological modification in adaptation to changed conditions of life. If an animal so exceedingly intolerant of fresh water as is a marine jellyfish may yet have all its tissues changed so as to adapt them to thrive in fresh water, and even die after an exposure of one minute to their ancestral element, assuredly we can see no reason why any animal in earth or sea or anywhere else may not in time become fitted to change its element." *

* While these sheets are passing through the press, a paper has been read before the Royal Society by Mr. A. G. Bourne, describing the hydroid stage of the fresh-water Medusa (*Proc. Roy. Soc.*, Dec. 11, 1884). He has discovered the hydroids on the roots of the *Pontederia*, which have been growing in the Lily-tank for several years, and which are therefore probably the source from which the tank became impregnated with the Medusæ.

CHAPTER X.

STAR-FISH AND SEA-URCHINS.

Structure of Star-fish and Sea-Urchins.

WE shall now proceed to consider in the organization of the Echinodermata a type of nervous system which is more highly developed than that of the Medusæ. In conducting this research, I was joined by my friend Professor J. Cossar Ewart, to whose unusual skill and untiring patience the anatomical part of the inquiry is due. But here, as formerly, I shall devote myself to the physiology of the subject, as it is not possible within the limits assigned to this volume to travel further into morphology than is necessary for the purpose of rendering the experiments intelligible. I shall therefore begin by seeking to give merely such a general idea of the structure of the Echinodermata as is necessary for this purpose.

As we all know, a Star-fish consists of a central disc and five radiating arms (Fig. 32). Upon the whole of the upper surface there occur numerous calcareous nodules embedded in the soft flesh, and supporting short spines. One of these nodules is much larger than any of the others, is constant in position, and is called the madreporic tubercle (Fig 32, m). Continuing our examination of the upper surface, we may observe, when we use a lens, a number of small pincer-like organs scattered about between the calcareous nodules, or attached to the spines; these are known as the pedicellariæ. Each

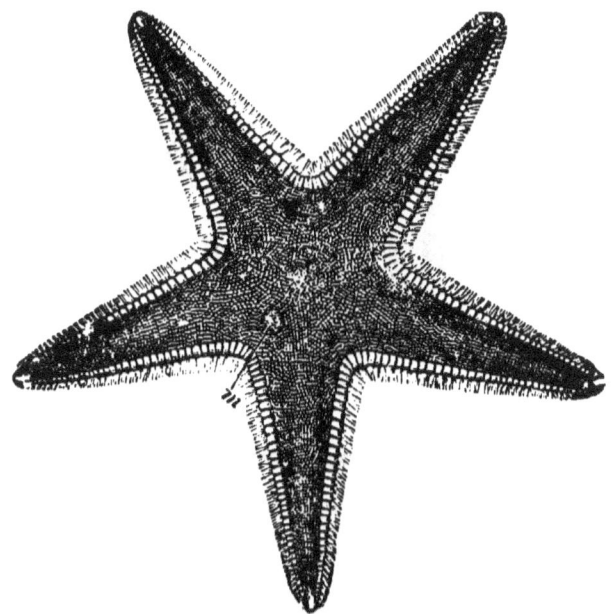

Fig. 32. Upper surface of a Star-fish (*Astropecten*). (From Cassell's " Nat. Hist.")

consists of a stalk serving to support a pair of forceps or pincers, and the whole being provided with muscles, the stalk is able to sway about and the pincers to open and shut (Fig. 33). The entire mechanism is therefore clearly adapted to seizing and holding on to something; but what it is that these curious organs are thus adapted to seize, and therefore of what use they are in the economy of the animal, has long been a standing puzzle to naturalists. I hope presently to be able to show that we have succeeded in doing something towards the solution of this puzzle.

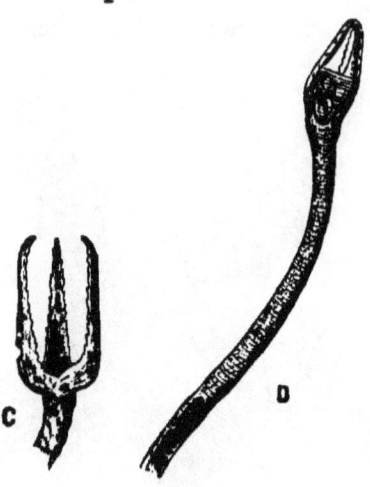

Fig. 33.—Pedicellariæ (magnified). (From Cassell's "Nat. Hist.")

Turning now to the under surface of our Star-fish (Fig. 34), we observe that the mouth is situated in the centre of the disc, and that from this mouth as a centre there radiate five grooves or furrows, which severally extend to the tips of each of the five rays. On each side of these grooves there are numerous actively moving membraneous tubes, which may be protruded or retracted by being filled or emptied with fluid. These are used for crawling, and I shall therefore call them the feet, or pedicels.

So much, then, for the external surface of a Star-fish. If, now, we examine the internal structure, we find that the central mouth leads by a short œsophagus into a central stomach, and that this in turn communicates with the intestine, which terminates in an orifice on the dorsal surface. Springing from the intestine at its origin, there are five tubes, each of which divides into two, and the five pairs of tubes thus formed extend into the five rays;

Fig. 34.—Lower surface of common Star-fish.

numerous blind processes grow out from these tubes, and give rise to glandular structures, which probably perform the functions of a liver.

When a section is made across the base of one of the arms, the furrows or grooves before mentioned are seen to be formed of two rows of plates connected together so as to compose a series of structures not unlike the couples of an ordinary roof. These so-called ambulacral plates rest on horizontal spine bearing plates, from which other larger plates extend upwards to form the sides of the arms.

In a living Star-fish the tube-feet or pedicels already mentioned are seen projecting from each side of the ambulacral groove; and, with the exception of a few at the tip of each arm, all the tube-

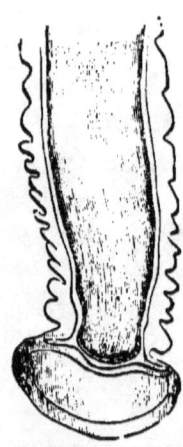

Fig. 35.—The terminal portion of a tube foot (magnified).

feet terminate in a well-formed sucker, by means of which they can be firmly fixed to a flat surface (Fig. 35).

If we wish to understand the structure and mechanism of this locomotor or ambulacral system —which, I may observe in passing, is of special interest from the fact that as a mechanism it is unique in the animal kingdom—we must resort to dissection. We then find that each of the tube-feet is provided in its membranous walls with a number of annular or ring-shaped muscular fibres; when these fibres contract, the fluid contained in the tube is forced back, while, conversely, when these fibres relax, the fluid runs into the tube. If the contraction of these fibres is strong, the tube shrinks up entirely, *i.e.* is retracted within the body of the animal; but if the contraction of the fibres is not so strong, the tube is only shortened. If, before its shortening, its terminal expansion, or sucker, has been applied to any flat surface, the effect of the shortening is to cause the sucker to adhere to the flat surface, in consequence of the pressure of the surrounding sea-water being greater than that of the fluid within the shortened tube. In this way, by alternately contracting and relaxing the muscular fibres in the walls of a tube-foot, a Star-fish is able alternately to cause the terminal sucker to fasten upon and to leave go of any flat surface upon which the animal may be crawling. In other words, when the tube-foot is about to form its attachment to a flat surface, it is fully distended with fluid; but when the terminal sucker touches the flat surface, this fluid is partly withdrawn, so causing the sucker to adhere.

When we dissect out one of these tube-feet, we find that at its base, within the body of the animal, it bifurcates into two branches. One of these branches passes immediately into a closed sac (Fig. 36, *f*), while the other passes into a large tube (Fig. 36, *k*), which runs all the way from base to tip of the ray, receiving in its course similar branches from all the

260 JELLY-FISH, STAR-FISH, AND SEA-URCHINS.

tube-feet in the ray. This common or radial tube itself opens into a circular tube (Fig. 36, *e*) surrounding the mouth of the animal (Fig. 36 *m*). This circular tube therefore receives five radial tubes—one from each of the five rays—and is likewise in communication with a number of membraneous sacs (Fig. 36, *c*, *d*), resembling in their structure (though

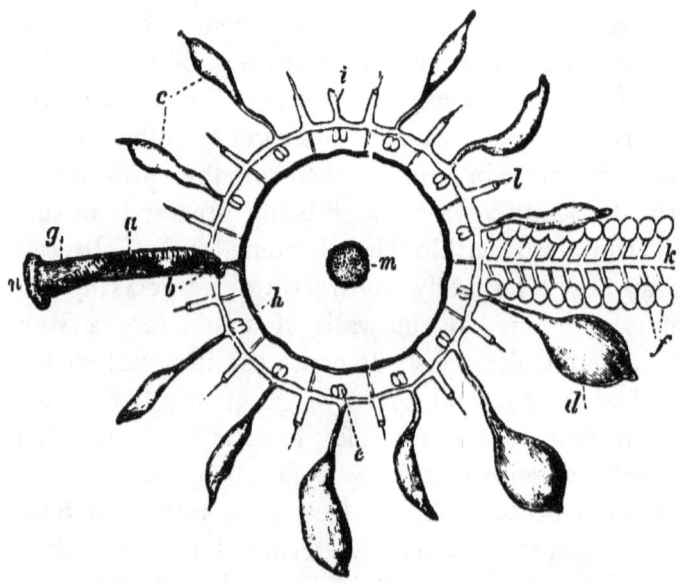

Fig. 36.—Diagram of ambulacral system of a Star-fish; *a*, madreporic canal; *b*, inner end; *g*, outer end of sinus leading to circular neural vessel; *h*, from which radial neural vessels, *l*, arise; *c d*, Polian vesicles; *f*, ampullæ; *m*, oral aperture; *n*. madreporic plate.

larger in size) those which occur at the base of each of the tube-feet. The function both of these sacs and of those at the base of each tube-foot is the same, namely, that of acting as reservoirs of the fluid when this is expelled from the tube-feet.

Moreover, all these membraneous sacs are provided with ring-shaped muscular fibres in their membraneous walls, which therefore serve as antagonists to the ring-shaped muscles which occur in the membraneous walls of the tube-feet; that is to say, when the muscles of the reservoirs contract (Fig. 36, c, d, f,), the pressure in the tube-feet is increased, and when these muscles relax, that pressure is diminished. The animal is thus furnished with the means of varying the head of pressure in its tube-feet, either locally or universally.

The circular tube surrounding the mouth communicates at one point with a calcareous tube (Fig. 36, a), which runs straight to the dorsal surface of the animal, and there terminates in the madreporic tubercle, to which I have already directed attention (Fig. 32, m, and Fig. 36, m). Thus it will be seen that all the pedicels of all the rays are in communication, by means of a closed system of tubes, with this madreporic tubercle. It has therefore been surmised that the function of this tubercle is that of acting as a filter to the sea-water which in large part constitutes the fluid that fills the ambulacral system. We have been able to prove that this surmise is correct; for we found that if we injected any part of the ambulacral system with coloured fluid—maintaining the injection for several hours at as great a pressure as the tubes would stand without rupturing—the coloured fluid found its way up the calcareous tube to the madreporic tubercle, on arriving at which it slowly oozed through the porous substance of which that tubercle consists.

Such, then, is the so-called ambulacral system of the Star-fish. Passing over another system of vessels which I need not wait to describe (Fig. 36, *g, h, l*), we come next to the nervous system. This is disposed on a very simple plan. It consists of a pentagonal ring surrounding the mouth, from which a nerve-trunk passes into each of the five rays, to run along the ambulacral groove as far as the extreme tip of the ray, where it ends in a small red pigmented spot, about which I shall have more to say presently. Each of these five radial nerves gives off in its course a number of delicate branches to the tube-feet.

Modifications of the Star-fish Type.

So much, then, for the structure of the common Star-fish. I must next say a few words on the remarkable modifications which this structure undergoes in different members of the Star-fish group.

In some species the size of the central disc is increased so as to fill up the interspaces between the rays, the whole animal being thus converted into the form of a pentagon. In other species, again, the reverse process has taken place, the rays having become relatively longer, and being at the same time very active; they look like five little snakes joined together by a circular disc (Fig. 37). Again, in another species the rays have begun to branch, these branches again to branch, and so on till the whole animal looks like a mat. But the most extreme modifications are attained in the sea cucumbers and lily-stars (Fig. 38). Without, however, waiting to consider these, I shall go a little more particularly into the modification of Star-fish structure which is presented by the sea-urchin, or Echinus (Fig. 39).

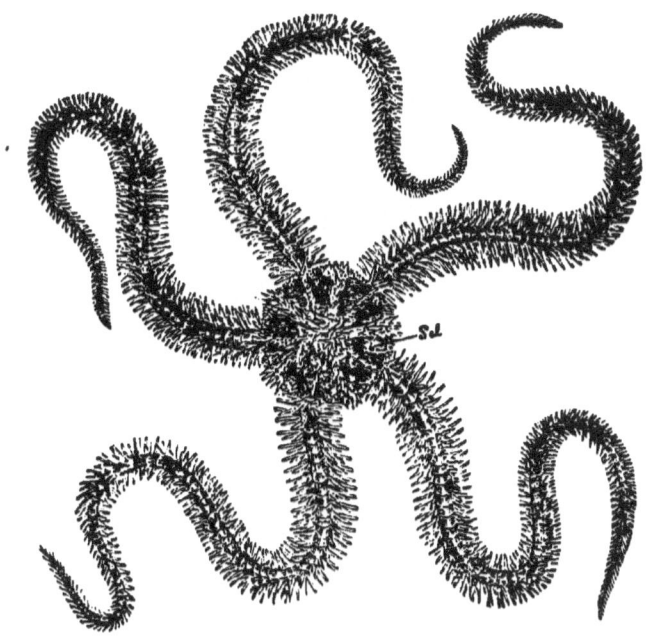

Fig 37.—A Brittle-star. (From Cassell's "Nat. Hist.")

Externally, the animal presents the form of an orange, and is completely covered with a large number of hard calcareous spines, on which account it derives its scientific name of Echinus, or hedgehog (the spines have been removed from the larger portion of the specimen represented in Fig. 39). In the living animal these spines are fully movable in all directions, each being mounted on a ball-and-

264   JELLY-FISH, STAR-FISH, AND SEA-URCHINS.

socket joint, and provided with muscles at its base. On the external surface, besides the spines, we

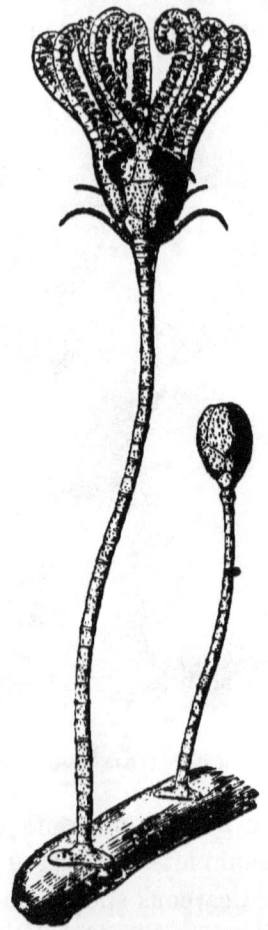

Fig. 38.—A Lily-star. (From Cassell's "Nat. Hist.")

meet with pedicellariæ (Fig. 33 magnified), and also with the madreporic tubercle (Fig. 39, *m*). The pedicellariæ in their main features resemble those which occur in the Star-fish, though considerably larger, and the ambulacral system is constructed

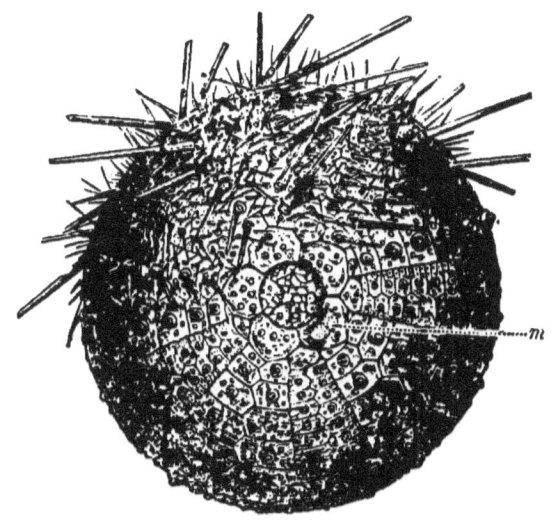

Fig. 39.—An Echinus, partly denuded of its spines. (From Cassell's "Nat. Hist.")

upon the same plan. If we shave off the spines and pedicellariæ (Fig. 39), we find that we come to

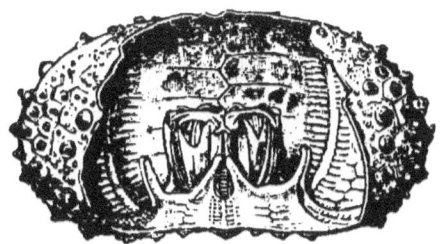

Fig. 40.—Showing interior of Echinus shell. (From Cassell's "Nat. Hist.")

a hard shell, which, if we break, we find to be hollow and filled with fluid (Fig. 40). The fluid closely resembles sea-water, but is, nevertheless, richly corpusculated; it coagulates when exposed to the air, and otherwise shows that it is something more than mere sea-water. If we look closely into the shell which has been deprived of its spines, we find that it is composed of a great number of small hexagonal plates (Fig. 41), the edges of which fit so closely together that the whole shell is converted into a box, which, when the animal is alive, is water-tight, as we have proved by submitting the contained fluid to hydrostatic pressure, under which

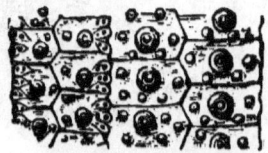

Fig. 41.—A portion of the external shell of an Echinus denuded of spines and slightly magnified, showing the arrangement of the plates, the balls in the ball-and-socket joints of the spines, and the holes through which the ambulacral feet are protruded. (From Cassell's "Nat. Hist.")

circumstances there is no leakage until the pressure is sufficient to burst the shell. Nevertheless, if we look closely at the dried shell of an Echinus, we shall see that it is not an absolutely closed box; for we shall see that the hexagonal plates are so arranged as to give rise to five double rows of holes or pores (Fig. 41), which extend symmetrically from pole to pole of the animal (Fig. 39). It is through these holes that the tube-feet are protruded; so that if we imagine a pentagonal species of Star-fish to be curved into the shape of a hollow spheroid, and then converted into a calcareous box with holes left for its feet to come through, we should have a mental picture of an Echinus. It would only be necessary to add the curious apparatus of teeth (Figs. 40 and 42), which occurs in the Echinus, to increase the size of the spines and pedicellariæ, and to make a few other such minor alterations; but in all its main features an Echinus is merely a Star-fish with its five rays calcified and soldered together so as to constitute a rigid box.

This echinoid type itself varies considerably among its numerous constituent species as to size, shape, length and thickness of the spines, etc.; but

Fig. 42.—Teeth of Echinus (from Cassell's "Nat. Hist.")

I need not wait to go into these details. Again, merely inviting momentary attention to the developmental history of these animals, I may remark that the phases of development through which an individual Echinoderm passes are not less varied and remarkable than are the permanent forms eventually assumed by the sundry species.

Natural Movements.

Turning now to the physiology of the Star-fish group, I shall begin by describing the natural movements of the animals.

Taking the common Star-fish as our starting-point, I have already explained the mechanism of its ambulacral system. The animals usually crawl in a determinate direction, and when in the course of their advance the terminal feet of the advancing ray—which are used, not as suckers, but as feelers, protruded forwards—happen to come into contact with a solid body, the Star-fish may either continue its direction of advance unchanged, or may turn towards the body which it has touched. Thus, for instance, while crawling along the floor of a tank, if the terminal feet at the end of a ray happen to touch a perpendicular side of the tank, the animal may either at once proceed to ascend this perpendicular side, or it may continue its progress along the floor, feeling the perpendicular side with the end of its rays perhaps the whole way round the tank, and yet not choosing, as it were, to ascend. In the cases where it does ascend and reaches the surface of the water, a Star-fish very often performs a number of peculiar movements, which we may call acrobatic (Fig. 43). On reaching the surface, the animal does not wish to leave its native element—in fact, cannot do so, because its sucking feet can only act under water—and neither does it wish again to descend into the levels from which it has just ascended. It, therefore, begins to feel about for rocks or sea-weeds at the surface, by crawling along the side of the tank, and every now and then throwing back its uppermost ray or rays along the surface of the water to feel for any solid support that may be within reach. If it finds one, it may very likely attach its uppermost rays to it, and then, letting go its other attachments, swing from the one support to the other. The activity and co-ordination manifested in these acrobatic movements are surprising, and give to the animal an almost intelligent appearance.

In Astropecten the ambulacral feet have become partly rudimentary, inasmuch as they have lost

Fig. 43.—Natural movements of a Star-fish on reaching the surface of water.

their terminal suckers (Fig. 44). These Star-fish, therefore, assist themselves in locomotion by the muscular movements of their rays, while they use their suckerless feet to run along the ground somewhat after the manner of centipedes. It is to be noticed, however, that although the feet have lost their suckers, the Star-fish is still able to make them adhere to solid surfaces in a comparatively inefficient manner, by constricting the tube on one side after it has brought this side into opposition with the solid surface (Fig. 45).

In the Brittle-stars the ambulacral feet have been still more reduced to rudiments, and are of no use at all, either as suckers or for assisting in locomotion. These Star-fish have, therefore, adopted another method of locomotion, and one which is a great

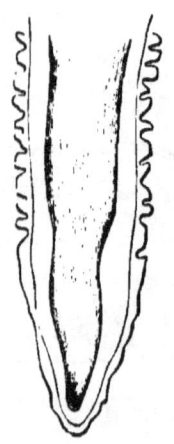 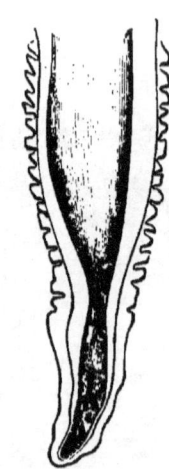

Fig. 44.—A pedicel of Astropecten (magnified), showing the absence of any terminal sucker.

Fig. 45.—The same, showing the method of extemporizing a sucker.

improvement upon the slow crawling of other members of the Star-fish group. The rays of the Brittle-stars are very long, flexible, and muscular, and by their combined action the animal is able to shuffle along flat horizontal surfaces. When it desires to move rapidly, it uses two of its opposite arms upon the horizontal floor with a motion like swimming (Fig. 46); at each stroke the animal advances with a leap or bound about the distance of two inches, and as the strokes follow one another rapidly, the Star-fish is able to travel at the rate of six feet per minute. A common Star-fish, on the other hand, with its slow crawling method of

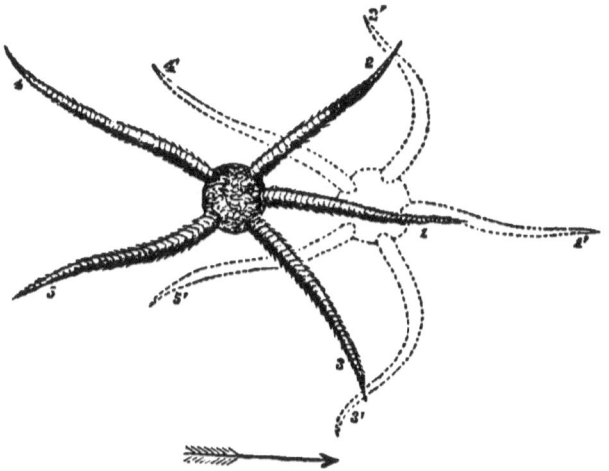

Fig. 46.—Natural movements of a Brittle-star when proceeding along a solid horizontal surface.

progression, can only go two inches per minute. Some of the Comatulæ, in which the muscularity of the rays has proceeded still further, are able actually to swim in the water by the co-ordinated movements of their rays.*

---

* In this case the locomotion of a Star-fish comes to be performed on the same plan or method as that of a Jelly-fish—the five rays performing, by their co-ordinated action, the same function as a swimming-bell. It is a curiously interesting fact, that although no two plans or mechanisms of locomotion could well be imagined as more fundamentally distinct than those which are respectively characteristic of these two groups of

The Echinus crawls in the same way as the common Star-fish; but besides its long suckers it also uses its spines, which by their co-ordinated action push the animal along. The suckers, moreover, in being protruded from all sides of a globe instead of from the under side of a flat organism, are of much more use as feelers than they are in the Star-fish. Therefore, while advancing, the feet facing the direction of advance are always kept extended to their fullest length, in order to feel for any object which the animal may possibly be approaching. When a perpendicular surface is reached, the Echinus may either ascend it or not, as in the case of the Star-fish. While walking, the animal keeps pretty persistently in one direction of advance. If it be partly rotated by the hand, it does not continue in the same direction, but continues its own movements as before; so that, for instance, if it is turned half round, it will proceed in a direction opposite to that in which it had previously been going. When at rest, some of the feet are used as anchors, and others protruded as feelers.

Regarded from the standpoint of the evolutionist, we have here an interesting series of gradations. At one end of the series we have the Echinus with its rays all united in a box-like rigid shell. At the other end of the series we have the Brittle-stars and Comatulæ with their highly muscular and

animals, nevertheless in this particular case and in virtue of special modification, a Star-fish should have adopted the plan or mechanism of a Jelly-fish.

mobile rays. Midway in the series we have Astropecten and the common Star-fish, where the rays are flexible and mobile, though not nearly so much so as in the Brittle-stars. Now, the point to observe is, that in correlation with this graduated difference in the mobility of the rays, there is a correspondingly graduated difference in the development of the ambulacral system of suckers. For in Echinus this system is seen in its most elaborate and efficient form; in the common Star-fish the suckers are still the most important organs of locomotion, though the muscularity of the rays has begun to tell upon the development of the specially ambulacral system, the suckers not being so long or so powerful as they are in Echinus. Lastly, the Brittle-stars and Comatulæ have altogether discarded the use of their sucking feet in favour of the much more efficient organs of locomotion supplied by their muscular rays; and, as a consequence, their feet have dwindled into useless rudiments, while the rays have become limb-like in their activity.

There is only one other point in connection with the natural movements of the Echinodermata which it is necessary for me to touch upon. All the species when turned upon their backs are able again to right themselves; but seeing, as I have just observed, that the organs of locomotion in the different species are not the same, the methods to which these species have to resort in executing the righting manœuvre are correspondingly diverse. Thus, the Brittle-stars can easily perform the needful manœuvre by wriggling some of their snakelike arms under the inverted disc, and heaving the whole body over by the mere muscularity of these organs. The common Star-fish, however, experiences more difficulty, and executes the manœuvre mainly by means of its suckers. That is to say, it twists round the tip of one or more of its rays (Fig. 47) until the ambulacral feet there situated are able to

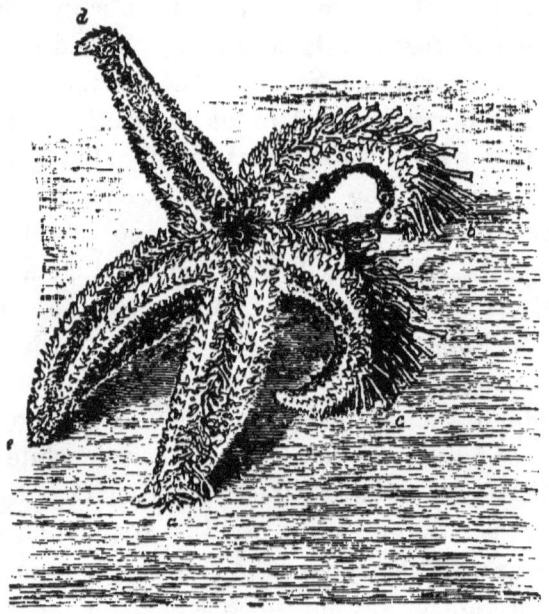

Fig. 47.—Natural righting movements of common Star-fish.

get a firm hold of the floor of the tank (*a*); then, by a successive and similar action of the ambulacral feet further back in the series, the whole ray is twisted round (*b*), so that the ambulacral surface of the end is applied flat against the floor of the tank (*c*). The manœuvre continuing, the semi-turn or spiral travels progressing all the way down the ray. Usually two or three adjacent rays perform this manœuvre simultaneously; but if, as sometimes happens, two opposite rays should begin to do so, one of them soon ceases to continue the manœuvre, and one or both of the rays adjacent to the other takes it up instead, so assisting and not thwarting the action. The spirals of the co-operating rays being invariably turned in the same direction (Fig. 47, *a*, *b*, and *c*), the result is, when they have proceeded sufficiently far down the rays, to drag over the remaining rays, which then abandon their hold of the bottom of the tank, so as not to offer any resistance to the lifting action of the active rays. The whole movement does not occupy more than half a minute. As a general rule, the rays are from the first co-ordinated to effect the righting movement in the direction in which it is finally to take place—the rays which are to be the active ones alone twisting over, and so twisting that all their spirals turn in the same direction.

A Star-fish (Astropecten) which is intermediate between the Brittle-star and the common Star-fish, in that its ambulacral feet are partly aborted (having lost their suckers, as shown in Fig. 44) and its rays more mobile than those of the common Star-fish, rights itself after the manner shown in Fig. 48, where the animal is represented as standing on the tips of four of its rays, while the fifth one is just about to be thrown upwards and over the others, in order to carry with it the two adjacent rays, and so eventually to overbalance the system round the fulcrum supplied by the tips of the other two rays, and thus bring the animal down upon its ventral surface.

But it is in the case of Echinus that these righting movements become most interesting, from the fact that they are so much more difficult to accom-

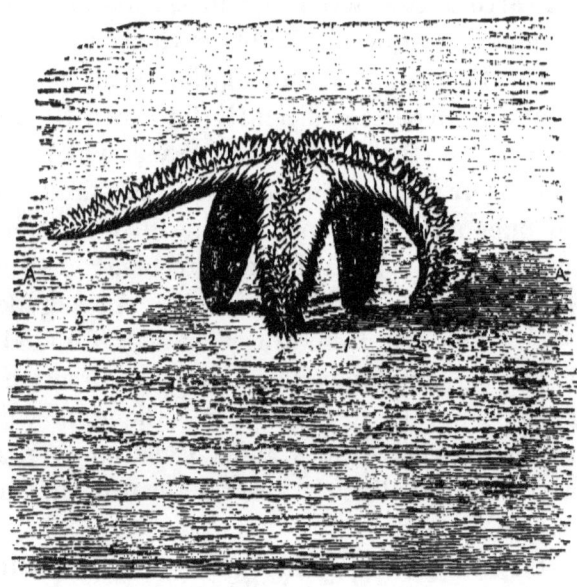

Fig 48.—Righting movements of Astropecten.

plish than they are in the case of the Star-fishes. For while a Star-fish is provided with flat, flexible, and muscular rays, comprising a small and light mass in relation to the motive power, an Echinus is a rigid, non-muscular, and globular mass, whose only motive power available for conducting the manœuvre is that which is supplied by its relatively feeble ambulacral feet. It is, therefore, scarcely surprising that unless the specimens chosen for these observations are perfectly fresh and vigorous, they are unable to right themselves at all; they remain permanently inverted till they die. But if the specimens are fresh and vigorous, they are sooner or later sure to succeed in righting themselves, and their method of doing so is always the same. Two, or perhaps three, adjacent rows of suckers are chosen out of the five, as the rows which are to accomplish the task (Fig. 49). As

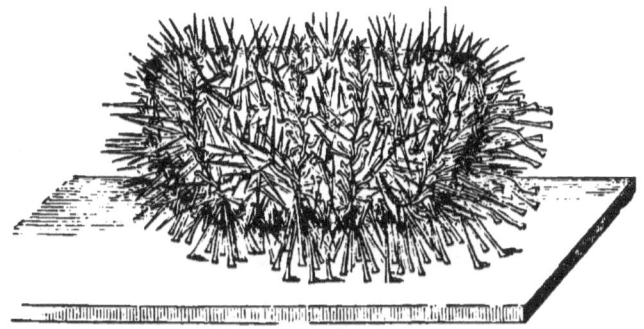

Fig. 49.

many feet upon the rows as can reach the floor of the tank are protruded downwards and fastened firmly to the floor; their combined action then serves to tilt the globe slightly over in their own direction, the anchoring feet on the other or opposite rows meanwhile releasing their hold of the tank to admit of this tilting (Fig. 50). The effect of this tilting is to enable the next feet in the active ambulacral rows to touch the floor of the tank, and, when they have established their hold,

278   JELLY-FISH, STAR-FISH, AND SEA-URCHINS.

they assist in increasing the tilt; then the next feet in the series lay hold, and so on, till the globe

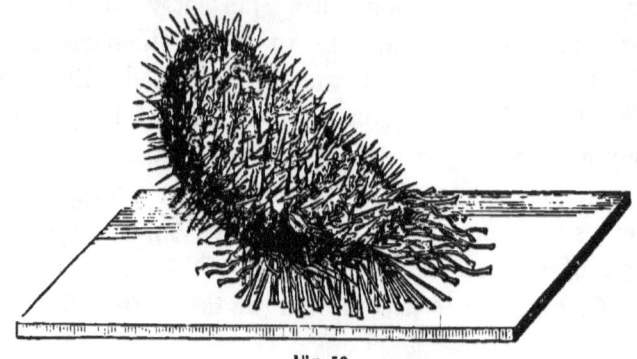

Fig. 50.

slowly but steadily rises upon its equator (Fig. 51). The difficulty of raising such a heavy mass into

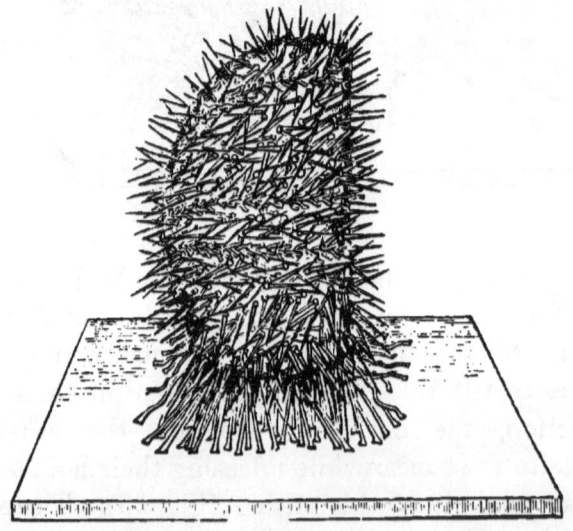

Fig. 51.

this position by means of the slender motive power

available can be at once appreciated on witnessing
the performance, so that one is surprised, notwithstanding the co-ordination displayed by all the
suckers, that they are able to accomplish the work
assigned to them. That the process is in truth a
very laborious one is manifest, not only from the
extreme slowness with which it takes place, but
also because, as already observed, in the case of not
perfectly strong specimens complete failure may
attend the efforts to reach the position of resting on
the equator—the Echinus, after rearing up a certain
height, becoming exhausted and again falling back
upon its ab-oral pole. Moreover, in some cases it
is interesting to observe that when the equator
position has been reached with difficulty, the
Echinus, as it were, gives itself a breathing space
before beginning the movement of descent—drawing
in all its pedicels save those which hold it securely
in the position to which it has attained, and
remaining in a state of absolute quiescence for a
prolonged time. It then suddenly begins to protrude
all its feet again, and to continue its manœuvre.
At any time during such a period of rest, a stimulus
of any kind will immediately determine a recom-
mencement of the manœuvre.

It will be perceived that as soon as the position
just described has been attained, gravity, which had
hitherto been acting in opposition to the righting
movement, now begins to favour that movement.
It might, therefore, be anticipated that the Echinus
would now simply let go all its attachments and
allow itself to roll over into its natural position

280   JELLY-FISH, STAR-FISH, AND SEA-URCHINS.

But an Echinus will never let go its attachments without some urgent reason, seeming to be above all things afraid of being rolled about at the mercy of currents; and therefore in this case it lets itself down almost as slowly as it raised itself up. So gently, indeed, is the downward movement effected, that an observer can scarcely tell the precise moment at which the righting is concluded. Therefore, in the downward movement, the feet, which at the earlier part of the manœuvre were employed

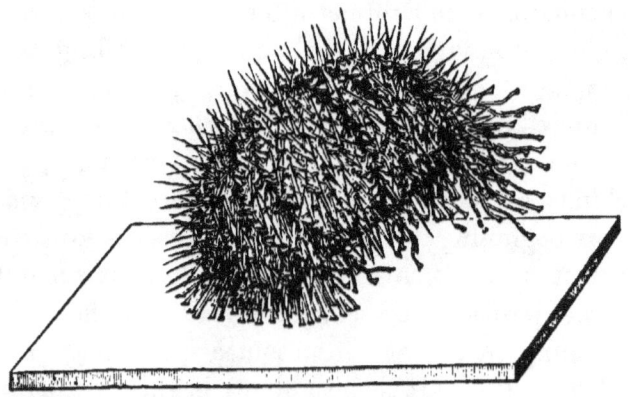

Fig. 52.

successfully in rearing the globe upon its equator, are now employed successfully in preventing its too rapid descent (Fig. 52).

Several interesting questions arise with reference to these righting movements of Echinus. First of all we are inclined to ask what it is that determines the choice of the rows of feet which are delegated to effect the movements. As the animal has a geometrical form of perfect symmetry, we might suppose that when it is placed upon its pole, all the five rows of feet would act in antagonism to one another; for there seems nothing more to determine either the action or the inaction of one row rather than another. Indeed, if there were any moral philosophers among the Echinoderms, they might point with triumph to the fact of their being able to right themselves as an irrefutable argument in favour of the freedom of the Echinoderm will. "We are in form," they might say, "perfectly geometrical, and our feet-rows are all arranged with perfect symmetry; therefore there is no reason, apart from the sovereign freedom of our choice, why we should ever use one set of feet rather than another in executing this important movement." And indeed, I do not see how these Echinoderm philosophers could be answered by any of the human philosophers, who, with less mathematical data and with less physiological reason, employ analogous arguments to prove the freedom of the human will. Physiologists, however, would give these Echinoderm philosophers the same answer that they are in the habit of giving to the human philosophers, viz. that although the physiological conditions are very nicely balanced, they are never *so* nicely balanced as to leave positively nothing to determine which rows of feet —that is to say, which sets of nerves—shall be used. And in this connection I may observe that on making a number of trials it becomes apparent in the case of certain individual specimens that they manifested a marked tendency to rotate always in the same direction, or to use the same set of foot-rows for the purpose of righting themselves.

In these individual specimens, therefore, we must conclude that the foot-rows thus employed are selected because of some slight accidental prepotency or superiority over the others; the animal has, as it were, thus much individual *character* as the result of a slight prepotency of some of its nerve-centres over the others.

Another question of still more interest arises out of these righting movements, namely, that as to their prompting cause. This question, however, I shall defer till later on, since it cannot be answered without the aid of experiments as distinguished from observation.

Stimulation.

In now quitting our observations on the natural movements of the Echinodermata, and beginning an account of the various experiments which we have tried upon these animals, I shall first take the experiments in stimulation.

All the Echinodermata seek to escape from injury. Thus, for instance, if a Star-fish or an Echinus is advancing continuously in one direction, and if it be pricked or otherwise irritated on any part of an excitable surface facing the direction of advance, the animal immediately reverses that direction. There is one point of special interest concerning these movements of response to stimulation. The form of the animals and the distribution of the nervous system being, as I have before said, of geometrical regularity, it follows that by applying two stimuli simultaneously on two different aspects of the animal, the combined result of these two stimuli is that of furnishing a very pretty instance in physiology of the physical principle of the parallelogram of forces. Thus, for instance, if two stimuli of equal intensity be applied simultaneously at the opposite sides of a globular Echinus, the animal begins to walk in a direction at right angles to an imaginary line joining these two points. And, generally, wherever the two points of simultaneous stimulation may be situated, the direction of the animal's advance is the diagonal between them. As showing in more detail how very delicate is the physiological balancing of stimuli which may be produced in these organisms, and consequently the manner in which we are able to play, as it were, upon their geometrically disposed nervous systems in illustration of the mechanical principle of the composition of forces, I shall quote a series of observations.

"1. Scraped with a scalpel the equator of an Echinus at two points opposite to each other—animal crawled at right angles to the line of injury.

"2. Similarly scraped at the ab-oral pole—no effect. There was no reason why injury here should determine escape in one direction rather than in another.

"3. Scraped similarly near the oral pole, and half-way between pole and equator—little or no effect.

"4. Scraped in rapid succession five equatorial and equidistant injuries—Echinus crawled actively in one determinate direction; the equal and equidistant injuries all round the globe neutralized one another.

"5. Scraped a band of uniform width all the way round the equator—same result as in 4.

"6. Band of injury in same specimen was then widened in the side facing the direction of crawling —no effect. Still further widened—slight change of direction, and, after a time, persistent crawling away from the widest part of the injured zone. Repeated this experiment on other specimens by scraping round the whole equator, and simultaneously making one part of the zone of injury wider than the rest—same result; the animal crawled away from the *greatest amount* of injury.

"7. Scraped on one side of the equator, and, after the animal had been crawling in a direct line from the source of irritation for a few minutes, similarly scraped equator on the opposite side—animal reversed its direction of crawling; it crawled away from the stimulus *supplied latest.*

"8. Scraped a number of places on all aspects of the animal indiscriminately—direction of advance uncertain and discontinuous, with a strong tendency to rotation upon vertical axis."

These observations show conclusively that the whole external surface, not only of the soft and fleshy Star-fish, but even of the hard and rigid Echinus, is everywhere sensitive to stimulation. Closer observation shows that this sensitiveness, besides being so general, is highly delicate. For if any part of the external surface of an Echinus is lightly touched with the point of a needle, all the feet, spines, and pedicellariæ within reach of that part, and even beyond it, immediately converge and close in upon the needle, grasp it, and hold it fast. This simultaneous movement of such a little forest of prehensile organs is a very beautiful spectacle to witness. In executing it the pedicellariæ are the most active, the spines somewhat slower, and the feet very much slower. The area affected is usually about half a square inch, although the pedicellariæ even far beyond this area may bend over towards the seat of stimulation, which, however, from their small size they are not able to reach.

And here we have proof of the function of the pedicellariæ—proof which we consider to be important, because, as I have before said, the use of these organs has so long been a puzzle to naturalists. In climbing perpendicular or inclined surfaces of rock, covered with waving sea-weeds, it must be of no small advantage to an Echinus to be provided on all sides with a multitude of forceps, all mounted on movable stalks, which instantaneously bring their grasping forceps to bear upon and to seize a passing frond. The frond being thus arrested, the spines come to the assistance of the pedicellariæ, and both together hold the Echinus to the support furnished by the sea-weed. Moreover the sea-weed is thus held steady till the ambulacral feet have time also to establish their hold upon it with their sucking discs. That the grasping and arresting of fronds of sea-weed in this way for the purposes of locomotion constitute an important function of the pedicellariæ, may at once be rendered evident

experimentally by drawing a piece of sea-weed over
the surface of a healthy Echinus in the water. The
moment the sea-weed touches the surface of the
animal, it is seen and felt to be seized by a number
of these little grasping organs, and—unless torn
away by a greater force than is likely to occur in
currents below the surface of the sea—it is held
steady till the ambulacral suckers have time to
establish their attachments upon it. Thus there is
no doubt that the pedicellariæ are able efficiently
to perform the function which we regard as
their chief function. We so regard this function,
not merely because it is the one that we observe
these organs chiefly to perform, but also because
we find that their whole physiology is adapted
to its performance. Thus their multitudinous
number and ubiquitous situation all over the
external surface of the animal is suggestive of
their being adapted to catch something which may
come upon them from any side, and which may
have strings and edges so fine as to admit of being
enclosed by the forceps. Again, the instantaneous
activity with which they all close round and seize
a moving body of a size that admits of their seizing
it, is suggestive of the objects which they are
adapted to seize being objects which rapidly brush
over the surface of the shell, and therefore objects
which, if they are to be seized at all, must be seized
instantaneously. Lastly, we find, on experimenting upon pedicellariæ, whether *in situ* or when
separated from the Echinus, that the clasping action
of the forceps is precisely adapted to the function

which we are considering ; for not only is the force exerted by the forceps during their contraction of an astonishing amount for the size of the organ (the serrated mandibles of the trident pedicellariæ holding on with a tenacity that can only have reference to some objects liable to be dragged away from their grasp), but it is very suggestive that this wonderfully tenacious hold is spontaneously relaxed after a minute or two. This is to say, the pedicellariæ tightly fix the object which they have caught for a time sufficient to enable the ambulacral suckers to establish their connections with it, and then they spontaneously leave go ; their grasp is not only so exceedingly powerful while it lasts, but it is as a rule timed to suit the requirements of the pedicels.*

Concerning the physiology of the pedicellariæ little further remains to be said. It may be stated, however, that the mandibles, which are constantly

---

* A further proof that this is at least one of the functions of the pedicellariæ is furnished by a simple experiment. If an Echinus is allowed to attach its feet to a glass plate held just above its ab-oral pole, and this plate be then raised in the water so that the Echinus is freely suspended in the water by means of its feet alone, the animal feels, as it were, that its anchorage is insecure, and actively moves about its unattached feet to seek for other solid surfaces. Under such circumstances it may be observed that the pedicellariæ also become active, and especially so near the surface of attachment, as if seeking for pieces of sea-weed. If a piece is presented to them, they lay hold upon it with vigour.

Of course the pedicellariæ may also have other functions to perform, and in a Star-fish Mr. Sladen has seen them engaged in cleaning the surface of the animal ; but we cannot doubt that at least in Echinus their main function is that which we have stated.

swaying about upon their contractile stalks as if in search for something to catch, will snap at an object only if it touches the inner surface of one or more of the expanded mandibles. Moreover, in the larger pedicellariæ, a certain part of the inner surface of the mandibles is much more sensitive to contact than is the rest of that surface; this part is a little pad about one-third of the way down the mandible: a delicate touch with a hair upon this part of any of the three mandibles is certain to determine an immediate closure of all the three. It is obvious that there is an advantage in the sensitive area, or zone, being placed thus low enough down in the length of the mandibles to ensure that the whole apparatus will not close upon an object till the latter is far enough within the grasp of the mechanism to give this mechanism the best possible hold. If, for instance, the tips of the mandibles were the most sensitive parts, or even if their whole inner surfaces were uniformly sensitive, the apparatus would be constantly closing upon objects when these merely brushed past their tips, and therefore closing prematurely for the purpose of grasping. But, as it is, the apparatus is admirably adapted to waiting for the best possible chance of getting a secure hold, and then snapping upon the object with all the quickness and tenacity of a spring-trap.

Another point worth mentioning is that if, after closure, any one or more of the mandibles be gently stroked on its outer surface near the base, all the mandibles are by this stimulation usually, though not invariably, induced again to expand. This is the only part of the whole organ the stimulation of which thus exerts an inhibitory influence on the contractile mechanism. If there is any functional purpose served by such relaxing influence of stimulating this particular part of the apparatus, we think it can only be as follows. When a portion of sea-weed brushes this particular part, it must be well below the tips of the mandibles, and therefore in a position where it, or some over-lying portion, may soon pass between the mandibles, if the latter are open; hence when touched in this place the mandibles, if closed, open to receive the sea-weed, should any part of it come within their cavity.

Turning next to experiments in stimulation with reference to the spines, I may observe that we have found these organs to be, physiologically considered, highly remarkable and interesting, from the fact that they display co-ordinated action in a degree which entitles them to be regarded as a vast multitude of limbs. Thus, for instance, if an Echinus be taken out of the water and placed upon a table, it is no longer able to use its feet for the purpose of locomotion, as their suckers are only adapted to be used under water. Yet the animal is able to progress slowly by means of the co-ordinated action of its spines, which are used to prop and push the globe-like shell along in some continuous direction. If, while the animal is thus slowly progressing, a lighted match be held near it, facing the direction of advance, as soon as the animal comes close enough to feel the heat, all the spines begin to make the animal move away in the opposite direction. Moreover, as showing the high degree in which the action of the spines is co-ordinated, I may mention that there is an urchin-like form of Echinoderm, which is called Spatangus, and which differs from the Echinus in having shorter feet and longer spines. When, therefore, a Spatangus is inverted, it is unable to right itself by means of its feet, as these are too short to admit of being used for this purpose; but, nevertheless, the animal is able to right itself by means of the co-ordinated action of its long spines, these being used successively and laboriously to prop and push the animal over in some one definite direction. The process takes a very long time to accomplish, and there are generally numerous failures, but the creature perseveres until it eventually succeeds.

Coming now to stimulation with reference to the feet, we find that when a drop of acid, or other severe stimulation, is applied to any part of a row of protruded pedicels, the entire row is immediately retracted, the pedicels retracting successively from the seat of irritation—so that if the latter be in the middle point of the series, two series of retractions are started, proceeding in opposite directions simultaneously; the rate at which they travel is rather slow. This process of retraction, however, although so complete within the ray irritated, does not extend to the other rays. But if the stimulus be applied to the centre of the disc, upon the oral surface of the animal, all the feet in all the rays are more or less retracted—the process of retraction radiating serially from the centre of stimulation. The influence of the stimulus, however, diminishes perceptibly with the distance from the centre. Thus, if weak acid be used as the irritant, it is only the feet near the bases of the rays that are retracted; and even if very strong acid be so used, it is only the feet as far as one-half or two-thirds of the way up the rays that are fully retracted—the remainder only having their activity impaired, while those near the tip may not be affected at all. If the drop of acid be placed on the dorsal, instead of the ventral surface of the disc, the effect on the feet is found to be just the converse; that is, the stimulus here applied greatly increases the activity of the feet. Further experiments show that this effect is produced by a stimulus applied anywhere over the dorsal aspect of the animal; so that, for instance, if a drop of acid be placed on the skin at the edge of a ray, and therefore just external to the row of ambulacral feet, the latter will be stimulated into increased activity; whereas, if the drop of acid had been placed a very small distance past the edge of the ray, so as to touch some of the feet themselves, then the whole row would have been drawn in. We have here rather an interesting case of antagonism, which is particularly well marked in Astropecten, on account of the active writhing movements which the feet exhibit when stimulated by an irritant placed on the dorsal surface of the animal. It may be added that in this antagonism the inhibitory function is the stronger; for when the feet are in active motion, owing to an irritant acting on the dorsal surface, they may be reduced to immediate quiescence—*i.e.* retracted—by placing another irritant on the ventral surface of the disc. Similarly, if retraction has been produced by placing the irritant on the ventral surface of the disc, activity cannot be again induced by placing another drop of the irritant on the dorsal surface.

Now, if we regard all these facts of stimulation taken together, it becomes evident that the external organs of an Echinoderm—feet, spines, and pedicellariæ—are all highly co-ordinated in their action; and therefore the probability arises that they are all held in communication with one another by means of an external nervous plexus. Accordingly we set to work on the external surface of the Echinus to see whether we could obtain any evidence of such a plexus microscopically. This we succeeded in doing, and afterwards found that Professor Lovèn had already briefly mentioned such a plexus as having been observed by him. The plexus consists of cells and fibres, closely distributed all over the surface of the shell, immediately under the epidermal layer of cells (Figs. 53, 54, 55), and it sends fibres all the way up the feet, spines, and pedicellariæ. As it seemed to us important to investigate the physiological properties of this plexus, Professor Ewart and I made a number of further experiments, an account of which will now lead us on to the next division of our subject, or that of section.

STAR-FISH AND SEA-URCHINS.

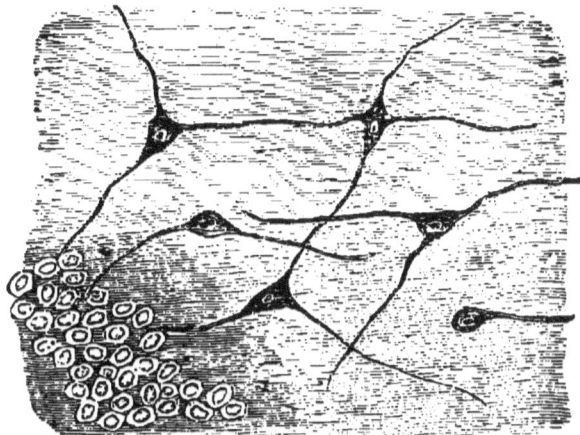

Fig. 53. External nerve-plexus of Echinus.

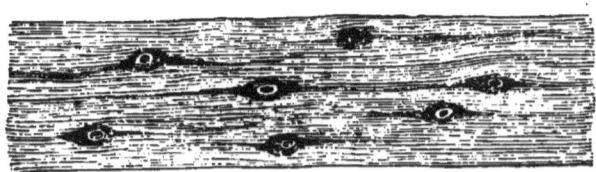

Fig. 54. Structure of a nerve-trunk of Echinus.

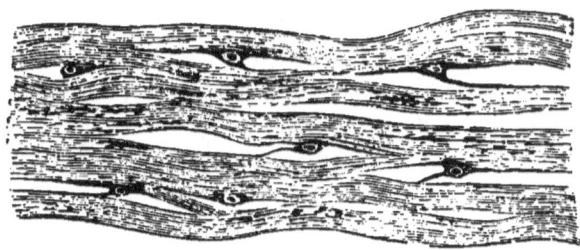

Fig. 55. Nerve-cells lying among the muscular fibres at the base of a spine in Echinus.

Section.

1. *Star-fish.*—Single rays detached from the organism crawl as fast and in as determinate a direction as do the entire animals. They also crawl up perpendicular surfaces, and sometimes away from injuries; but they do not invariably, or even generally, seek to escape from the latter, as is so certain to be the case with entire animals. Lastly, when inverted, separated rays right themselves as quickly as do the unmutilated organisms.

Dividing the nerve in any part of its length has the effect, whether or not the ray is detached from the animal, of completely destroying all physiological continuity between the pedicels on either side of the line of division. Thus, for instance, if the nerve be cut across half-way up its length, the row of pedicels is at once physiologically bisected, one-half of the row becoming as independent of the other half as it would were the whole ray divided into two parts: that is to say, the distal half of the row may crawl while the proximal half is retracted, or *vice versâ;* and if a drop of acid be placed on either half, the serial contraction of the pedicels in that half stops abruptly at the line of nerve-division. As a result of this complete physiological severance, when a detached ray so mutilated is inverted, it experiences much greater difficulty in righting itself than it does before the nerve is divided. The line of nerve-injury lies flat upon the floor of the tank, while the central and distal portions of the ray, *i.e.* the portions on either side of that line, assume various movements and shapes. The central portion is particularly apt to take on the form of an arch, in which the central end of the severed ray and the line of nerve-section constitute the points of support (tetanus ?) (Fig. 56), or the central end may from the first show paralysis, from which it never recovers. The distal end, on the other hand, usually continues active, twisting about in various directions, and eventually fastening its tip upon the floor of the tank to begin the spiral movement of righting itself. This movement

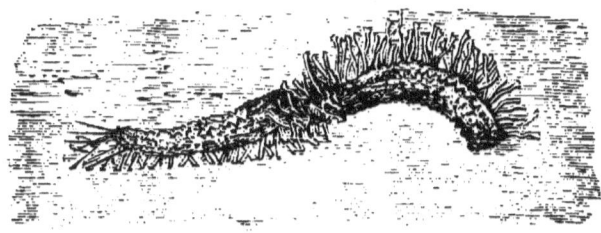

Fig. 56. Movements performed by a detached ray of a Star-fish, in which the central nerve-trunk is divided.

then continues as far as the line of nerve-injury, where it invariably stops (Fig. 56). The central portion may then be dragged over into the normal position, or may remain permanently inverted, according to the strength of pull exerted by the distal portion; as a rule, it does not itself assist in the righting movement, although its feet usually continue protruded and mobile. Thus, the effect of a transverse section of the nerve in a ray is that of completely destroying physiological continuity between the pedicels on either side of the section.

The only other experiments in nerve-section to which the simple anatomy of a Star-fish exposes itself is that of dividing the nerve-ring in the disc; or, which is virtually the same thing, while leaving this intact, dividing all the nerves where they pass from it into the rays. In specimens mutilated by severing the nerves at the base of each of the five rays, or by dividing the nerve-ring between all the rays, the animal loses all power of co-ordination among its rays. When a common Star-fish is so mutilated it does not crawl in the same determinate manner as an unmutilated animal, but, if it moves at all, it moves slowly and in various directions. When inverted, the power of effecting the righting manœuvre is seen to be gravely impaired, although eventually success is always achieved. There is a marked tendency, as compared with unmutilated specimens, to a promiscuous distribution of spirals and doublings, so that instead of a definite plan of the manœuvre being formed from the first, as is usually the case with unmutilated specimens, such a plan is never formed at all; among the five rays there is a continual change of un-coördinated movements, so that the righting seems to be eventually effected by a mere accidental prepotency of some of the righting movements over others. Appended is a sketch of such un-coördinated movement, taken from a specimen which for more than an hour had been twisting its rays in various directions (Fig. 57). Another sketch is appended to show a form of bending which specimens mutilated as described are very apt to manifest, especially just after the operation. When placed upon their dorsal surface, they turn up all their rays with a peculiar and exactly similar curve in each, which gives to the animal a somewhat tulip-like form (Fig. 58). This form is never assumed by unmutilated specimens, and in mutilated ones, although it may last for a long time, it is never permanent. In detached rays

Fig. 57. Un-coördinated movements of a Star-fish, in which the nerves of all the rays have been divided.

this peculiar curve is also frequently exhibited; but if the nerve of such a ray is divided at any point in its length, the curve is restricted to the distal portion of the ray, and it stops abruptly at the line of nerve-section. When entire Star-fish are mutilated by a section of each nerve-trunk half-way up each ray, and the animal is then placed upon its back, the tetanic contraction of the muscles in the rays

298   JELLY-FISH, STAR-FISH, AND SEA-URCHINS.

before mentioned as occurring under this form of section in detached rays, has the effect, when now occurring in all the rays, of elevating the disc from the floor of the tank. This opisthotonous-like spasm is not, however, permanent; and the distal ends of the rays forming adhesions to the floor of the tank, the animal eventually rights itself, though

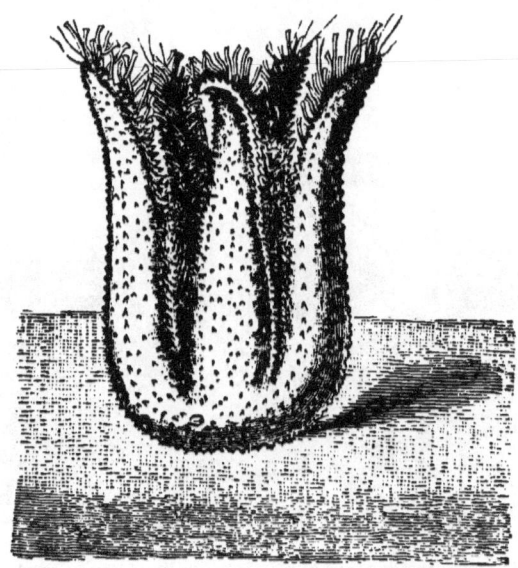

Fig. 58. Form frequently assumed by Star-fish under similar circumstances.

much more slowly than unmutilated specimens. After it has righted itself, although it twists about the distal portions of the rays, it does not begin to crawl for a long time, and when it does so, it crawls in a slow and indeterminate manner. Star-fish so mutilated, however, can ascend perpendicular surfaces.

The loss of co-ordination between the rays caused by division of the nerve-ring in the disc is rendered most conspicuous in Brittle-stars, from the circumstance that in locomotion and in righting so much here depends upon co-ordinated muscular contraction of the rays. Thus, for instance, when a Brittle-star has its nerve-ring severed between each ray, an interesting series of events follows. First, there is a long period of profound shock—spontaneity, and even irritability, being almost suspended, and the rays appearing to be rigid, as if in tetanic spasm. After a time, feeble spontaneity returns—the animal, however, not moving in any determinate direction. Irritability also returns, but only for the rays immediately irritated, stimulation of one ray causing active writhing movements in that ray, but not affecting, or only feebly affecting, the other rays. The animal, therefore, is quite unable to escape from the source of irritation, the aimless movements of the rays now forming a very marked contrast to the instantaneous and vigorous leaping movements of escape which are manifested by unmutilated specimens. Moreover, unmutilated specimens will vigorously leap away, not only from stimulation of the rays, but also from that of the disc; but those with their nerve-ring cut make no attempts to escape, even from the most violent stimulation of the disc. In other words, the disc is entirely severed from all physiological connection with the rays.

If the nerve-ring be divided at two points, one on either side of a ray, that ray becomes physiologically separated from the rest of the organism. If the two nerve-divisions are so placed as to include two adjacent rays—*i.e.* if one cut is on one side of a ray and the other on the further side of an adjacent ray—then these two rays remain in physiological continuity with one another, although they suffer physiological separation from the other three. When a Brittle-star is completely divided into two portions, one portion having two arms and the other three, both portions begin actively to turn over on their backs, again upon their faces, again upon their backs, and so on alternately for an indefinite number of times. These movements arise from the rays, under the influence of stimulation caused by the section, seeking to perform their natural movements of leaping, which however end, on account of the weight of the other rays being absent, in turning themselves over. An entire Brittle-star when placed on its back after division of its nerve-ring is not able to right itself, owing to the destruction of co-ordination among its rays. Astropecten, under similar circumstances, at first bends its rays about in various ways, with a preponderant disposition to the tulip form, and keeps its ambulacral feet in active movement. But after half an hour, or an hour, the feet generally become retracted and the rays nearly motionless—the animal, like a Brittle-star, remaining permanently on its back. In this, as in other species, the effect of dividing the nerve-ring on either side of a ray is that of destroying its physiological connection with the rest of the animal, the feet in that ray, although still remaining feebly active, no longer taking part in any co-ordinated movement—that ray, therefore, being merely dragged along by the others.

Under this division it only remains further to be said, that section of the nerve-ring in the disc, or the nerve-trunks of the rays, although, as we have seen, so completely destroying physiological continuity in the rows of ambulacral feet and muscular system of the animal, does not destroy physiological continuity in the external nerve-plexus; for however much the nerve-ring and nerve-trunks may be injured, stimulation of the dorsal surface of the animal throws all the ambulacral feet and all the muscular system of the rays into active movement. This fact proves that the ambulacral feet and the muscles are all held in nervous connection with one another by the external plexus, without reference to the integrity of the main nerve-trunks.

2. *Echini.—Section of external surface of shell.* —If a cork-borer be applied to the external surface of the shell of an Echinus, and rotated there till the calcareous substance of the shell is reached, and therefore a continuous circular section of the over-lying tissues effected, it is invariably found that the spines and pedicellariæ within the circular area are physiologically separated from the contiguous spines and pedicellariæ, as regards local reflex excitability. That is to say, if any part of this circular area be stimulated, all the spines and pedicellariæ within that area immediately respond to the stimulation in the ordinary way; while none

of the spines or pedicellariæ surrounding the area
are affected. Similarly, if any part of the shell
external to the circumscribed area be stimulated,
the spines and pedicellariæ within that area are not
affected. These facts prove that the function which
is manifested by these appendages of localizing
and gathering round a seat of stimulation, is exclusively dependent upon the external nerve-plexus.
It is needless to add that in this experiment it does
not signify of what size or shape or by what means
the physiological island is made, so long as the
destruction of the nervous plexus by a closed curve
of injury is rendered complete. In order to ascertain whether, in the case of an unclosed curve of
injury, any irradiation of a stimulus would take
place round the ends of the curve, we made sundry
kinds of section. It is, however, needless to describe
these, for they all showed that, after injury of a
part of the plexus, there is no irradiation of the
stimulus round the ends of the injury. Thus, for
instance, if a short straight line of injury be made,
by drawing the point of a scalpel over the shell, say
along the equator of the animal, and if a stimulus
be afterwards applied on either side of that line,
even quite close to one of its ends, no effect will be
exerted on the spines or pedicellariæ on the other
side of the line. This complete inability of a
stimulus to escape round the ends of an injury,
forms a marked contrast to the almost unlimited
degree in which such escape takes place in the more
primitive nervous plexus of the Medusæ.

Although the nervous connections on which the

spines and pedicellariæ depend for their function of localizing and closing round a seat of stimulation are thus shown to be completely destroyed by injury of the external plexus, other nervous connections, upon which another function of the spines depends, are not in the smallest degree impaired by such injury. The other function to which I allude is that which brings about the general co-ordinated action of all the spines for the purposes of locomotion. That this function is not impaired by injury of the external plexus is proved by the fact that if the area within a closed line of injury on the surface of the shell be strongly irritated, all the spines over the whole surface begin to manifest their peculiar bristling movements, and by this co-ordinated action rapidly move the animal in a straight line of escape from the source of irritation; the injury to the external plexus, although completely separating the spines enclosed by it from their neighbouring spines as regards what may be called their local function of seizing the instrument of stimulation, nevertheless leaves them in undisturbed connection with all the other spines in the organism as regards what may be called their universal function of locomotion.

Evidently, therefore, this more universal function must depend upon some other set of nervous connections; and experiment shows that these are distributed over all the *internal* surface of the shell. Our mode of experimenting was to divide the animal into two hemispheres, remove all the internal organs of both hemispheres (these operations producing no impairment of any of the functions of the pedicels, spines, or pedicellariæ), and then to paint with strong acid the inside of the shell—completely washing out the acid after about a quarter of a minute's exposure. The results of a number of experiments conducted on this method may be thus epitomized :—

The effect of painting the back or inside of the shell with strong acid (*e.g.* pure HCl) is that of at first strongly stimulating the spines into bristling movements, and soon afterwards reducing them to a state of quiescence, in which they lie more or less flat, and in a peculiarly confused manner that closely resembles the appearance of corn when "laid" by the wind. The spines have now entirely lost both their spontaneity and their power of responding to a stimulus applied on the external surface of the shell—*i.e.* their local reflex excitability, or power of closing in upon a source of irritation. These effects may be produced over the whole external surface of the shell, by painting the whole of the internal surface; but if any part of the internal surface be left unpainted, the corresponding part of the external surface remains uninjured. Conversely, if all the internal surface be left unpainted except in certain lines or patches, it will only be corresponding lines and patches on the external surface that suffer injury. It makes no difference whether these lines or patches be painted in the course of the ambulacral feet, or anywhere in the inter-ambulacral spaces.

The above remarks, which have reference to the

spines, apply equally to the pedicellariæ, except that their spontaneity and reflex irritability are not destroyed, but only impaired.

Some hours after the operation it usually happens that the spontaneity and reflex irritability of the spines return, though in a feeble degree, and also those of the pedicellariæ, in a more marked degree. This applies especially to the reflex irritability of the pedicellariæ; for while their spontaneity does not return in full degree, their reflex irritability does—or almost in full degree.

These experiments, therefore, seem to point to the conclusions—1st, that the general co-ordination of the spines is dependent on the integrity of an internal nerve-plexus; 2nd, that the internal plexus is everywhere in intimate connection with the external; and 3rd, that complete destruction of the former, while profoundly influencing the functions of the latter, nevertheless does not wholly destroy them.

Professor Ewart therefore undertook carefully to examine the internal surface of the shell, to see whether any evidence of this internal nervous plexus could be found microscopically, and, after a great deal of trouble, he has succeeded in doing so. But as he has not yet published his results, I shall not forestall them further than to say that this internal plexus spreads all over the inside of the shell, and is everywhere in communication with the external plexus by means of fibres which pass between the sides of the hexagonal plates of which the shell of the animal is composed. Thus we can understand how it is that when a portion of the external plexus is isolated from the rest of that plexus as a result of the cork-borer experiment, the island still remains in communication with the nerve-centres which preside over the co-ordination of the spines, as proved by the fact of the Echinus using its spines to escape from irritation applied to the area included within the circle of injury to the external plexus produced by the cork-borer.

Now, where are these nerve-centres situated? We have just seen that we have evidence of the presence of such centres somewhere in an Echinus, seeing that all the spines exhibit such perfect co-ordination in their movements. Where, then, are these centres?

Seeing that in a Star-fish the rays are co-ordinated in their action by means of the pentagonal ring in the disc, analogy pointed to the nervous ring round the mouth of an Echinus as the part of the nervous system which most probably presides over the co-ordinated action of the spines. Accordingly, we tried the effect of removing this nervous ring, and immediately obtained conclusive proof that this was the centre of which we were in search; for as soon as the nervous ring was removed, the Echinus lost, completely and permanently, all power of co-ordination among its spines. That is to say, after this operation these organs were never again used by the animal for the purposes of locomotion, and no matter how severe an injury we applied, the Echinus, when placed on a table, did not seek to escape. But the spines were not wholly paralyzed, or motionless. On the contrary, their power of spontaneous movement continued unimpaired, as did also their power of closing round a seat of irritation on the external surface of the shell. The same remark applies to the pedicellariæ, and the explanation is simple. It is the external nervous plexus which holds all the spines and pedicellariæ in communication with one another as by a network; so that when any part of this network is irritated, all the spines and pedicellariæ in the neighbourhood move over to the seat of irritation. On the other hand, it is the internal plexus which serves to unite all the spines to the nerve-centre which surrounds the mouth, and which alone is competent to co-ordinate the action of all the spines for the purposes of locomotion.

It remains to consider whether the ambulacral feet exhibit any general co-ordinated action, and, if so, whether this likewise depends upon the same nerve-centre.

The fact already mentioned, that during progression an Echinus uses some of its feet for crawling and others for feeling its way, is enough to suggest that all the feet are co-ordinated by a nerve-centre. But in order to be quite sure about the fact of there being a general co-ordination among all the feet, we tried the following experiments.

I have already described the righting movements which are performed by an Echinus when the animal is inverted, and it will be remembered that in this animal the manœuvre is effected by means of the feet alone. At first sight this might almost seem sufficient to prove the fact of a general co-ordination among the feet; but further reflection will show that it is not so. For the feet being all arranged in regular series, when one row begins to effect the rotation of the globe, it may very well be that its further rotation in the same direction is due only to the fact that the slight tilt produced by the pulling of the first feet in the series A, B, C gives the next feet in the series D, E, F an opportunity of reaching the floor of the tank; their adhesions being established, they would tend by their pulling to increase still further the tilt of the globe, thus giving the next feet in the series an opportunity of fastening to the floor of the tank, and so on. In order, therefore, to see whether these righting movements were due to nervous co-ordination among the feet, or merely to the accident of the serial arrangement of the feet, we tried the experiments which I shall now detail.

First of all we took an Echinus, and by means of a thread suspended it upside-down in a tank of water half-way up the side of the tank, and in such a way that only the feet on one side of the ab-oral pole were able to reach the perpendicular wall of the tank. These feet as quickly as possible established their adhesions to the perpendicular wall, and, the thread being then removed, the Echinus was left sticking to the side of the tank in an inverted position by means of the ab-oral ends of two adjacent feet-rows (Fig. 59). Under these circumstances, as we should expect from the previous experiments, the animal sets about righting itself as quickly as possible. Now, if the righting action of the feet were entirely and only of a serial character, the righting would require to be performed by rearing the animal upwards; the effect of foot after foot in the same rows being applied in succession to the side of the tank, would require to be that of rotating the globular shell against the side of the tank towards the surface of the water, and therefore against the action of gravity. This is sometimes done, which proves that the energy required to perform the feat is not more than a healthy Echinus can expend. But much more frequently the Echinus adopts another device, and the only one by which it is possible for him to attain his purpose without the labour of rotating upwards: he rotates laterally and downwards in the form of a spiral. Thus, let us call the five feet-rows, 1, 2, 3, 4, and 5 (Figs. 59, 60, 61), and suppose that 1 and 2 are in use near their ab-oral ends in holding the animal inverted against the perpendicular side of a tank. The downward spiral rotation would then be effected by gradually releasing the outer feet in row 1, and simultaneously attaching the outer feet in row 2 (*i.e.* those nearest to row 3, and furthest from row 1), as far as possible to the outer side of that row. The effect of this is to make the globe roll far enough to that side to enable the inner feet of row 3 (*i.e.* those nearest to row 2), when fully protruded, to touch the side of the tank. They establish their adhesions, and the residue of feet in row 1, now leaving go their hold, these new adhesions serve to roll the globe still further round in the same

310   JELLY-FISH, STAR-FISH, AND SEA-URCHINS.

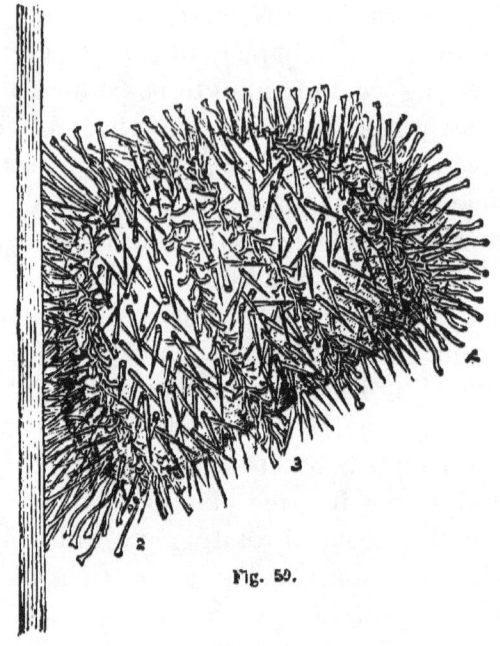

Fig. 59.

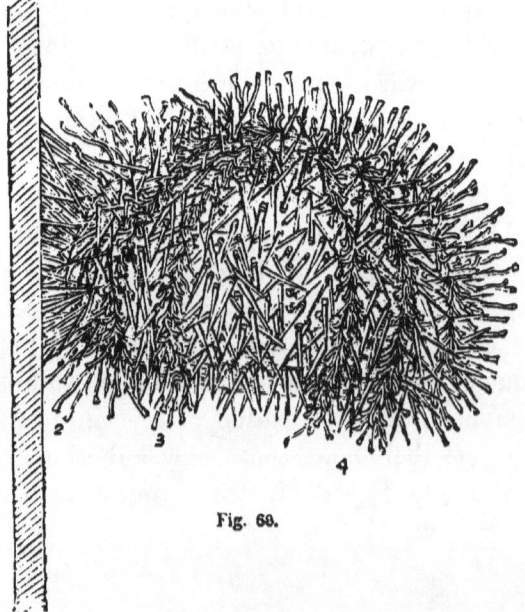

Fig. 60.

direction of lateral rotation, and so the process proceeds from row to row; but the globe does not merely roll along in a horizontal direction, or at the same level in the water, for each new row that comes into action takes care, so to speak, that the

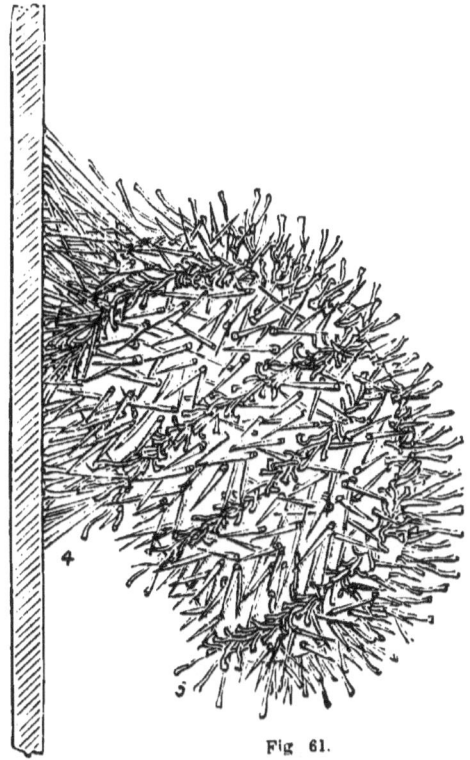

Fig 61.

Figs. 59, 60, and 61 are righting movements of Echinus on a perpendicular surface.

feet which it employs shall be those which are as far below the level of the feet in the row last employed as their length when fully protruded (i.e. their power of touching the tank) renders possible. The rotation of the globe thus becomes a double one, lateral and downwards, till the animal assumes its normal position with its oral pole against the perpendicular tank wall. So considerable is the rotation in the downward direction, that the normal position is generally attained before one complete lateral, or equatorial, rotation is completed.

The result of this experiment, therefore, implies that the righting movements are due to something more than the merely successive action of the series of feet to which the work of righting the animal may happen to be given. The same conclusion is pointed to by the results of the following experiment.

A number of vigorous Echini were thoroughly shaved with a scalpel over the whole half of one hemisphere, *i.e.* the half from the equator to the oral pole. They were then inverted on their ab-oral poles. The object of the experiment was to see what the Echini which were thus deprived of the lower half of three feet-rows would do when, in executing their righting manœuvres, they attained to the equatorial position and then found no feet wherewith to continue the manœuvre. The result of this experiment was first of all to show us that the Echini invariably chose the unmutilated feet-rows wherewith to right themselves. Probably this is to be explained, either by the general principle to which the escape from injury is due—viz. that injury inflicted on one side of an Echinoderm stimulates into increased activity the locomotor organs of the opposite side,—or by the consideration that destruction of the lower half of a row very probably induces some degree of shock in the remaining half, and so leaves the corresponding parts of the unmutilated rows prepotent over the mutilated one. Be this as it may, however, we found that the difficulty was easily overcome by tilting the animal over upon its mutilated feet-rows sufficiently far to prevent the unmutilated rows from reaching the floor of the tank. When held steadily in this position for a short time, the mutilated rows established their adhesions, and the Echinus was then left to itself. Under these circumstances an Echinus will always continue the manœuvre along the mutilated feet-rows with which it was begun, till the globe reaches the position of resting upon its equator, and therefore arrives at the line where the shaved area commences. The animal then remains for hours in this position, with a gradual but continuous motion backwards, which appears to be due to the successive slipping of the spines—these organs in the righting movements being always used as props for the ambulacral feet to pull against while rearing the globe to its equatorial position, and in performing this function on a slate floor the spines are liable often to slip. The only other motion exhibited by Echini thus situated is that of a slow rolling movement, now to one side and now to another, according to the prepotency of the pull exerted by this or that row of ambulacral feet. Things continue in this way until the slow backward movement happens to bring the animal against some side of the tank, when the uninjured rows of ambulacral feet immediately adhere to the surface and rotate the animal upwards or horizontally, until it attains the normal position. But if care be taken to prevent contact with any side of the tank, the mutilated Echinus will remain propped on its equator for days; it never adopts the simple expedient of reversing the action of its mutilated feet-rows, so as to bring the globe again upon its ab-oral pole and get its unmutilated feet-rows into action.

From this we may conclude that the righting movements of the pedicels are due, not to the merely serial action of the pedicels, but to their co-ordination by a nerve-centre acting under a stimulus supplied by a sense of gravity; for if the movements of the pedicels were merely of a serial character, we should not expect that the equatorial position, having been attained under these circumstances, should be permanently maintained. We should not expect this, because after a while the pedicels, which are engaged in maintaining the globe in its equatorial position, must become exhausted and relax their hold, when those next behind in the series would lay hold of the bottom of the tank, and so on, the rotation of the globe thus proceeding in the opposite direction to that in which it had previously taken place. On the other hand, if the righting movements of the pedicels are due to co-ordination proceeding from a nerve-centre acting under a sense of gravity, we should expect the animal under the circumstances mentioned to remain permanently reared upon its equator; for this would show that the nerve-centre was always persistently, though fruitlessly, endeavouring to co-ordinate the action of the absent feet.

Further, as proof that the ambulacral feet of Echinus are under the control of some centralizing apparatus when executing the righting manœuvre, we may state one other fact. When the righting macenuvre is nearly completed by the rows engaged in executing it, the lower feet in the other rows become strongly protruded and curved downwards, in anticipation of shortly coming into contact with the floor of the tank when the righting manœuvre shall have been completed (see Fig. 52, p. 280). This fact tends to show that all the ambulacral feet of the animal are, like all the spines, held in mutual communication with one another by some centralizing mechanism.

But the best proof of all that the feet in executing the righting manœuvre are under the influence of a co-ordinating centre, is one that arose from an experiment suggested to me by Mr. Francis Darwin, and which I shall now describe. Mr. Darwin having kindly sent the apparatus which his father and himself had used in their experiments on the geotropism of plants, it was employed thus. A healthy Echinus was placed in a large bottle filled to the brim with sea-water, and having been inverted on the bottom of the bottle, it was allowed in that position to establish its adhesions. The bottle was then corked and mounted on an upright wheel of the apparatus whereby, by means of clockwork, it could be kept in continual slow rotation in a vertical plane. The object of this was to ascertain whether the continuous rotation in a vertical plane would prevent the animal from righting itself (because confusing the nerve-centres which, under ordinary circumstances, could feel by their sense of gravity which was up and which was down), or would still allow the animal to right itself (because not interfering with the serial action of the feet). Well, it was found that this rotation of the whole animal in a vertical plane entirely prevented the righting movements during any length of time that it might be continued, and that these movements were immediately resumed as soon as the rotation was allowed to cease. This, moreover, was the case, no matter what phase of the righting manœuvre the Echinus might have reached at the moment when the rotation began. Thus, for instance, if the globe were allowed to have reached the position of resting on its equator before the rotation was commenced, the Echinus would remain motionless, holding on with its equatorial feet, so long as the rotation was kept up.

Therefore, there can be no question that the ambulacral feet are all under the influence of a co-ordinating nerve-centre, quite as much as are the spines. But, on the other hand, experiments show that the centre in this case is not of so localized a character as it is in the case of the spines; for when the nerve-ring is cut out, the co-ordination of the feet, although impaired, is not wholly destroyed. Take, for instance, the case of the righting manœuvre. The effect of cutting out the nerve-ring is that of entirely destroying the ability to perform this manœuvre in the case of the majority of specimens; nevertheless about one in ten continue able to perform it. Again, if an Echinus is divided into two hemispheres by an incision carried from pole to pole through any meridian, the two hemispheres will live for days, crawling about in the same manner as entire animals; if their ocular plates are not injured, they seek the light, and when inverted they right themselves. The same observations apply to smaller segments, and even to single detached rows of ambulacral feet. The latter are, of course, analogous to the single detached rays of a Star-fish, so far as the system of ambulacral feet is concerned; but, looking to the more complicated apparatus of locomotion (spines and pedicellariæ), as well as to the rigid consistence and awkward shape of the segment—standing erect, instead of lying flat—the appearance presented by such a segment in locomotion is much more curious, if not surprising, than that presented by the analogous part of a Star-fish under similar circumstances. It is still more surprising that such a fifth-part segment of an Echinus will, when propped up on its ab-oral pole (Fig. 62), right itself (Fig. 63) after the manner of larger segments or entire animals. They, however, experience more difficulty in doing so, and very often, or indeed generally, fail to complete the manœuvre.

On the whole, then, we may conclude that the nervous system of an Echinus consists (1) of an external plexus which serves to unite all the feet,

318   JELLY-FISH, STAR-FISH, AND SEA-URCHINS.

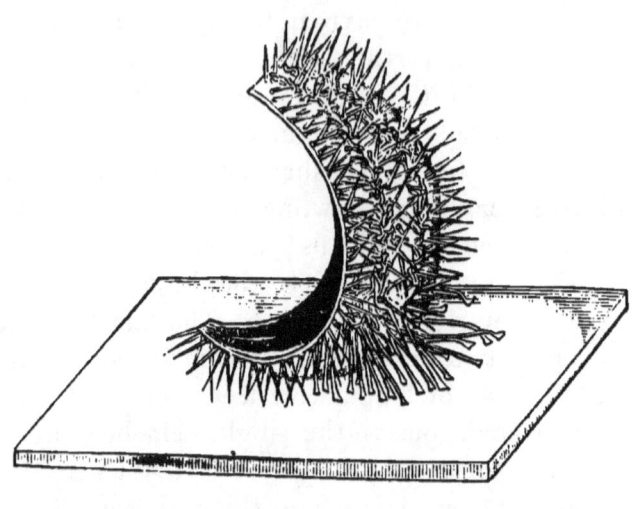

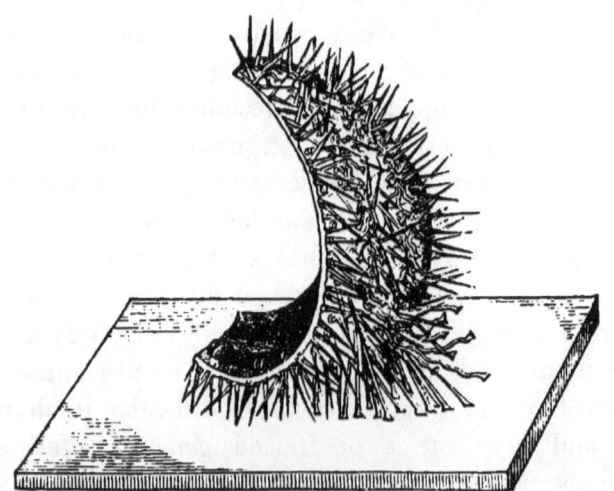

Figs. 62 and 63.—Righting and ambulacral movements of severed segments of Echinus.

spines, and pedicellariæ together, so that they all approximate a point of irritation situated anywhere in that plexus; (2) of an internal nervous plexus which is everywhere in communication through the thickness of the shell with the external, and the function of which is that of bringing the feet, spines, and probably also the pedicellariæ into relation with the great co-ordinating nerve-centre situated round the mouth; (3) of central nervous matter which is mainly gathered round the mouth, and there presides exclusively over the co-ordinated action of the spines, and in large part also over the co-ordinated action of the feet, but which is further in part distributed along the courses of the main nerve-trunks, and so secures co-ordination of feet even in separated segments of the animal.

Special Senses.

Before concluding, I must say a few words on the experiments whereby we sought to test for the presence in Echinoderms of the special senses of sight and smell.

We have found unequivocal evidence of the Star-fish (with the exception of the Brittle-stars) and the Echini manifesting a strong disposition to crawl towards, and remain in, the light. Thus, if a large tank be completely darkened, except at one end where a narrow slit of light is admitted, and if a number of Star-fish and Echini be scattered over the floor of the tank, in a few hours the whole number, with the exception of perhaps a few per cent., will be found congregated in the narrow slit of light. The source we used was diffused daylight, which was admitted through two sheets of glass, so that the thermal rays might be considered practically excluded. The *intensity* of the light which the Echinoderms are able to perceive may be very feeble indeed; for in our first experiments we boarded up the face of the tank with ordinary pine-wood, in order to exclude the light over all parts of the tank except at one narrow slit between two of the boards. On taking down the boards we found, indeed, the majority of the specimens in or near the slit of light; but we also found a number of other specimens gathering all the way along the glass face of the tank that was immediately behind the pine-boards. On repeating the experiment with blackened boards, this was never found to be the case; so there can be no doubt that in the first experiments the animals were attracted by the faint glimmer of the white boards, as illuminated by the very small amount of light scattered from the narrow slit through a tank, all the other sides of which were black slate. Indeed, towards the end of the tank, where some of the specimens were found, so feeble must have been the intensity of this glimmer, that we doubt whether even human eyes could have discerned it very distinctly. Owing to the prisms at our command not having sufficient dispersive power for the experiments, and not wishing to rely on the uncertain method of employing coloured glass, we were unable to ascertain how the Echinoderms might be affected by different rays.

On removing with a pointed scalpel the eye-spots from a number of Star-fish and Echini, without otherwise injuring the animals, the latter no longer crawled towards the light, even though this were admitted to the tank in abundance; but they crawled promiscuously in all directions. On the other hand, if only one out of the five eye-spots were left intact, the animals crawled towards the light as before. It may be added that single detached rays of Star-fish and fifth-part segments of Echini crawl towards the light in the same manner as entire animals, provided, of course, that the eye-spot is not injured.

The presence of a sense of smell in Star-fish was proved by keeping some of these animals for several days in a tank without food, and then presenting them with small pieces of shell-fish. The Star-fish immediately perceived the proximity of food, as shown by their immediately crawling towards it. Moreover, if a small piece of the food were held in a pair of forceps and gently withdrawn as the Star-fish approached it, the animal could be led about the floor of the tank in any direction, just as a hungry dog could be led about by continually withdrawing from his nose a piece of meat as he continually follows it up. This experiment, however, was only successful with Star-fish which had been kept fasting for several days; freshly caught Star-fish were not nearly so keen in their manifestations, and indeed in many cases did not notice the food at all.

Desiring to ascertain whether the sense of smell

were localized in any particular organs, as we had found to be the case with the sense of sight, I first tried the effect of removing the five ocelli. This produced no difference in the result of the above experiment with hungry Star-fish, and therefore I next tried the effect of cutting off the tips of the rays. The Star-fish behaving as before, I then progressively truncated the rays, and thus eventually found that the olfactory sense was equally distributed throughout their length. The question, however, still remained whether it was equally distributed over both the upper and the lower surfaces. I therefore tried the effect of varnishing the upper surface. The Star-fish continued to find its food as before, which showed that the sense of smell was distributed along the lower surface. I could not try the converse experiment of varnishing this surface, because I should thereby have hindered the action of the ambulacral feet. But by another method I was able nearly as well to show that the upper surface does not participate in smelling. This method consisted in placing a piece of shell-fish upon the upper surface and allowing it to rest there. When this was done, the Star-fish made no attempt to remove the morsel of food by brushing it off with the tips of its rays, as is the habit of the animal when any irritating substance is applied to this surface. Therefore I conclude that the upper or dorsal surface of a Star-fish takes no part in ministering to the sense of smell, which by the experiment of varnishing this surface, and also by that of progressively truncating the rays, is proved to be distributed over the whole of the ventral or lower surface of the animal. For I must add that severed rays behave in all these respects like the entire organisms, although they are disconnected from the mouth and disc.

As this chapter has already extended to so great a length, I omit from it any account of some further experiments which I tried concerning the effects of nerve-poisons upon the Echinodermata. A full record of these experiments may be found in the publications of the Linnean Society.

FINIS.

D. APPLETON & CO.'S PUBLICATIONS.

GEORGE J. ROMANES'S WORKS.

MENTAL EVOLUTION IN MAN: Origin of Human Faculty. One vol., 8vo. Cloth, $3.00.

This work, which follows "Mental Evolution in Animals," by the same author, considers the probable mode of genesis of the human mind from the mind of lower animals, and attempts to show that there is no distinction of kind between man and brute, but, on the contrary, that such distinctions as do exist all admit of being explained, with respect to their evolution, by adequate psychological analysis.

"The vast array of facts, and the sober and solid method of argument employed by Mr. Romanes, will prove, we think, a great gift to knowledge."— *Saturday Review.*

JELLY-FISH, STAR-FISH, AND SEA-URCHINS. Being a Research on Primitive Nervous Systems. 12mo. Cloth, $1.75.

"Although I have throughout kept in view the requirements of a general reader, I have also sought to render the book of service to the working physiologist, by bringing together in one consecutive account all the more important observations and results which have been yielded by this research."—*Extract from Preface.*

"A profound research into the laws of primitive nervous systems conducted by one of the ablest English investigators. Mr. Romanes set up a tent on the beach and examined his beautiful pets for six summers in succession. Such patient and loving work has borne its fruits in a monograph which leaves nothing to be said about jelly-fish, star-fish, and sea-urchins. Every one who has studied the lowest forms of life on the sea-shore admires these objects. But few have any idea of the exquisite delicacy of their structure and their nice adaptation to their place in nature. Mr. Romanes brings out the subtile beauties of the rudimentary organisms, and shows the resemblances they bear to the higher types of creation. His explanations are made more clear by a large number of illustrations."—*New York Journal of Commerce.*

ANIMAL INTELLIGENCE. 12mo. Cloth, $1.75.

"A collection of facts which, though it may merely amuse the unscientific reader, will be a real boon to the student of comparative psychology, for this is the first attempt to present systematically the well-assured results of observation on the mental life of animals."—*Saturday Review.*

MENTAL EVOLUTION IN ANIMALS. With a Posthumous Essay on Instinct, by CHARLES DARWIN. 12mo. Cloth, $2.00.

"Mr. Romanes has followed up his careful enumeration of the facts of 'Animal Intelligence,' contributed to the 'International Scientific Series,' with a work dealing with the successive stages at which the various mental phenomena appear in the scale of life. The present installment displays the same evidence of industry in collecting facts and caution in co-ordinating them by theory as the former."—*The Athenæum.*

New York: D. APPLETON & CO., 1, 3, & 5 Bond Street.

D. APPLETON & CO.'S PUBLICATIONS.

SIR JOHN LUBBOCK'S (Bart.) WORKS.

THE ORIGIN OF CIVILIZATION AND THE PRIMITIVE CONDITION OF MAN, MENTAL AND SOCIAL CONDITION OF SAVAGES. Fourth edition, with numerous Additions. With Illustrations. 8vo. Cloth, $5.00.

"This interesting work—for it is intensely so in its aim, scope, and the ability of its author—treats of what the scienti-ts denominate *anthropology*, or the natural history of the human species ; the complete science of man, body, and soul, including sex, temperament, race, civilization, etc."—*Providence Press*.

PREHISTORIC TIMES, AS ILLUSTRATED BY ANCIENT REMAINS AND THE MANNERS AND CUSTOMS OF MODERN SAVAGES. Illustrated. 8vo. Cloth, $5.00.

"This is, perhaps, the best summary of evidence now in our possession concerning the general character of prehistoric times. The Bronze Age, The Stone Age, The Tumuli, The Lake Inhabitants of Switzerland, The Shell Mounds, The Cave Man, and The Antiquity of Man, are the titles of the most important chapters."—*Dr. C. K. Adams's Manual of Historical Literature.*

ANTS, BEES, AND WASPS. A Record of Observations on the Habits of the Social Hymenoptera. With Colored Plates. 12mo. Cloth, $2.00.

"This volume contains the record of various experiments made with ants, bees, and wasps during the last ten years, with a view to test their mental condition and powers of sense. The author has carefully watched and marked particular insects, and has had their nests under observation for long periods—one of his ants' nests having been under constant inspection ever since 1874. His observations are made principally upon ants, because they show more power and flexibility of mind ; and the value of his studies is that they belong to the department of original research."

ON THE SENSES, INSTINCTS, AND INTELLIGENCE OF ANIMALS, WITH SPECIAL REFERENCE TO INSECTS. "International Scientific Series." With over One Hundred Illustrations. 12mo. Cloth, $1.75.

The author has here collected some of his recent observations on the senses and intelligence of animals, and especially of insects, and has attempted to give, very briefly, some idea of the organs of sense, commencing in each case with those of man himself.

THE PLEASURES OF LIFE. 12mo. Cloth, 50 cents ; paper, 25 cents.

CONTENTS.—THE DUTY OF HAPPINESS. THE HAPPINESS OF DUTY. A SONG OF BOOKS. THE CHOICE OF BOOKS. THE BLESSING OF FRIENDS. THE VALUE OF TIME. THE PLEASURES OF TRAVEL. THE PLEASURES OF HOME. SCIENCE. EDUCATION.

New York: D. APPLETON & CO., 1, 3, & 5 Bond Street.

D. APPLETON & CO.'S PUBLICATIONS.

JOHN TYNDALL'S WORKS.

ESSAYS ON THE FLOATING MATTER OF THE AIR, in Relation to Putrefaction and Infection. 12mo. Cloth, $1.50.

ON FORMS OF WATER, in Clouds, Rivers, Ice, and Glaciers. With 35 Illustrations. 12mo. Cloth, $1.50.

HEAT AS A MODE OF MOTION. New edition. 12mo. Cloth, $2.50.

ON SOUND: A Course of Eight Lectures delivered at the Royal Institution of Great Britain. Illustrated. 12mo. New edition. Cloth, $2.00.

FRAGMENTS OF SCIENCE FOR UNSCIENTIFIC PEOPLE. 12mo. New revised and enlarged edition. Cloth, $2.50.

LIGHT AND ELECTRICITY. 12mo. Cloth, $1.25.

LESSONS IN ELECTRICITY, 1875-'76. 12mo. Cloth, $1.00.

HOURS OF EXERCISE IN THE ALPS. With Illustrations. 12mo. Cloth, $2.00.

FARADAY AS A DISCOVERER. A Memoir. 12mo. Cloth, $1.00.

CONTRIBUTIONS TO MOLECULAR PHYSICS in the Domain of Radiant Heat. $5.00.

SIX LECTURES ON LIGHT. Delivered in America in 1872-'73. With an Appendix and numerous Illustrations. Cloth, $1.50.

FAREWELL BANQUET given at Delmonico's, New York. Paper, 50 cents.

ADDRESS delivered before the British Association, assembled at Belfast. Revised, with Additions. 12mo. Paper, 50 cents.

RESEARCHES ON DIAMAGNETISM AND MAGNE-CRYSTALLIC ACTION, including the Question of Diamagnetic Polarity. With Ten Plates. 12mo. Cloth, $1.50.

NEW FRAGMENTS. 12mo. Cloth, $2.00.

New York: D. APPLETON & CO., 1, 3, & 5 Bond Street.

D. APPLETON & CO.'S PUBLICATIONS.

CHARLES DARWIN'S WORKS.

ORIGIN OF SPECIES BY MEANS OF NATURAL SELECTION, OR THE PRESERVATION OF FAVORED RACES IN THE STRUGGLE FOR LIFE. From sixth and last London edition. 2 vols., 12mo. Cloth, $4.00.

DESCENT OF MAN, AND SELECTION IN RELATION TO SEX. With many Illustrations. A new edition. 12mo. Cloth, $3.00.

A NATURALIST'S VOYAGE AROUND THE WORLD. Journal of Researches into the Natural History and Geology of Countries visited during the Voyage of H. M. S. "Beagle." Illustrated with Maps and 100 Views of the places visited and described, chiefly from sketches taken on the spot, by R. T. PRITCHETT. 8vo. Cloth, $5.00.

Also popular edition. 12mo. Cloth, $2.00.

THE STRUCTURE AND DISTRIBUTION OF CORAL REEFS. Based on Observations made during the Voyage of the "Beagle." With Charts and Illustrations. 12mo. Cloth, $2.00.

GEOLOGICAL OBSERVATIONS on the Volcanic Islands and Parts of South America visited during the Voyage of the "Beagle." With Maps and Illustrations. 12mo. Cloth, $2.50.

EMOTIONAL EXPRESSIONS OF MAN AND THE LOWER ANIMALS. 12mo. Cloth, $3.50.

D. APPLETON & CO.'S PUBLICATIONS.

CHARLES DARWIN'S WORKS.—(*Continued.*)

THE VARIATIONS OF ANIMALS AND PLANTS UNDER DOMESTICATION. With a Preface, by Professor ASA GRAY. 2 vols. Illustrated. Cloth, $5.00.

INSECTIVOROUS PLANTS. 12mo. Cloth, $2.00.

MOVEMENTS AND HABITS OF CLIMBING PLANTS. With Illustrations. 12mo. Cloth, $1.25.

THE VARIOUS CONTRIVANCES BY WHICH ORCHIDS ARE FERTILIZED BY INSECTS. Revised edition, with Illustrations. 12mo. Cloth, $1.75.

THE EFFECTS OF CROSS AND SELF FERTILIZATION IN THE VEGETABLE KINGDOM. 12mo. Cloth, $2.00.

DIFFERENT FORMS OF FLOWERS ON PLANTS OF THE SAME SPECIES. With Illustrations. 12mo. Cloth, $1.50.

THE POWER OF MOVEMENT IN PLANTS. By CHARLES DARWIN, LL. D., F. R. S., assisted by FRANCIS DARWIN. With Illustrations. 12mo. Cloth, $2.00.

THE FORMATION OF VEGETABLE MOULD THROUGH THE ACTION OF WORMS. With Observations on their Habits. With Illustrations. 12mo. Cloth, $1.50.

New York: D. APPLETON & CO., 1, 3, & 5 Bond Street.

D. APPLETON & CO.'S PUBLICATIONS.

DR. HENRY MAUDSLEY'S WORKS.

BODY AND WILL: Being an Essay concerning Will in its Metaphysical, Physiological, and Pathological Aspects. 12mo. Cloth, $2.50.

BODY AND MIND: An Inquiry into their Connection and Mutual Influence, specially in reference to Mental Disorders. 1 vol., 12mo. Cloth, $1.50.

PHYSIOLOGY AND PATHOLOGY OF MIND:

PHYSIOLOGY OF THE MIND. New edition. 1 vol., 12mo Cloth, $2.00. CONTENTS: Chapter I. On the Method of the Study of the Mind.—II. The Mind and the Nervous System.—III. The Spinal Cord, or Tertiary Nervous Centres; or, Nervous Centres of Reflex Action.—IV. Secondary Nervous Centres; or, Sensory Ganglia; Sensorium Commune.—V. Hemispherical Ganglia; Cortical Cells of the Cerebral Hemispheres; Ideational Nervous Centres, Primary Nervous Centres; Intellectorium Commune.—VI. The Emotions.—VII. Volition.—VIII.—Motor Nervous Centres, or Motorium Commune and Actuation or Effection.—IX. Memory and Imagination.

PATHOLOGY OF THE MIND. Being the Third Edition of the Second Part of the "Physiology and Pathology of Mind," recast, enlarged, and rewritten. 1 vol., 12mo. Cloth, $2.00. CONTENTS: Chapter I. Sleep and Dreaming.—II. Hypnotism, Somnambulism, and Allied States.—III. The Causation and Prevention of Insanity: (A) Etiological.—IV. The same continued—V. The Causation and Prevention of Insanity: (B) Pathological.—VI. The Insanity of Early Life.—VII. The Symptomatology of Insanity.—VIII. The same continued.—IX. Clinical Groups of Mental Disease.—X. The Morbid Anatomy of Mental Derangement.—XI. The Treatment of Mental Disorders.

RESPONSIBILITY IN MENTAL DISEASE. (International Scientific Series.) 1 vol., 12mo. Cloth, $1.50.

"The author is at home in his subject, and presents his views in an almost singularly clear and satisfactory manner. . . . The volume is a valuable contribution to one of the most difficult and at the same time one of the most important subjects of investigation at the present day."— *New York Observer.*

"Handles the important topic with masterly power, and its suggestions are practical and of great value."—*Providence Press.*

New York: D. APPLETON & CO., 1, 3, & 5 Bond Street.

D. APPLETON & CO.'S PUBLICATIONS.

GENERAL PHYSIOLOGY OF MUSCLES AND NERVES. By Dr. I. ROSENTHAL, Professor of Physiology at the University of Erlangen. With seventy-five Woodcuts. ("International Scientific Series.") 12mo. Cloth, $1.50.

"The attempt at a connected account of the general physiology of muscles and nerves is, as far as I know, the first of its kind. The general data for this branch of science have been gained only within the past thirty years."—*Extract from Preface.*

SIGHT: An Exposition of the Principles of Monocular and Binocular Vision. By JOSEPH LE CONTE, LL. D., author of "Elements of Geology"; "Religion and Science"; and Professor of Geology and Natural History in the University of California. With numerous Illustrations. 12mo. Cloth, $1.50.

"It is pleasant to find an American book which can rank with the very best of foreign works on this subject. Professor Le Conte has long been known as an original investigator in this department; all that he gives us is treated with a master-hand."—*The Nation.*

ANIMAL LIFE, as affected by the Natural Conditions of Existence. By KARL SEMPER, Professor of the University of Würzburg. With 2 Maps and 106 Woodcuts, and Index. 12mo. Cloth, $2.00.

"This is in many respects one of the most interesting contributions to zoölogical literature which has appeared for some time."—*Nature.*

THE ATOMIC THEORY. By AD. WURTZ, Membre de l'Institut; Doyen Honoraire de la Faculté de Médecine; Professeur à la Faculté des Sciences de Paris. Translated by E. CLEMINSHAW, M. A., F. C. S., F. I. C., Assistant Master at Sherborne School. 12mo. Cloth, $1.50.

"There was need for a book like this, which discusses the atomic theory both in its historic evolution and in its present form. And perhaps no man of this age could have been selected so able to perform the task in a masterly way as the illustrious French chemist, Adolph Wurtz. It is impossible to convey to the reader, in a notice like this, any adequate idea of the scope, lucid instructiveness, and scientific interest of Professor Wurtz's book. The modern problems of chemistry, which are commonly so obscure from imperfect exposition, are here made wonderfully clear and attractive."—*The Popular Science Monthly.*

THE CRAYFISH. An Introduction to the Study of Zoölogy. By Professor T. H. HUXLEY, F. R. S. With 82 Illustrations. 12mo. Cloth, $1.75.

"Whoever will follow these pages, crayfish in hand, and will try to verify for himself the statements which they contain, will find himself brought face to face with all the great zoölogical questions which excite so lively an interest at the present day."

"The reader of this valuable monograph will lay it down with a feeling of wonder at the amount and variety of matter which has been got out of so seemingly slight and unpretending a subject."—*Saturday Review.*

New York: D. APPLETON & CO., 1, 3, & 5 Bond Street.

D. APPLETON & CO.'S PUBLICATIONS.

OUTINGS AT ODD TIMES. By CHARLES C. ABBOTT, author of "Days out of Doors" and "A Naturalist's Rambles about Home." 16mo. Cloth, gilt top, $1.25.

"A charming little volume, literally alone with Nature, for it discusses seasons and the fields, birds, etc., with the loving freedom of a naturalist born. Every page reads like a sylvan poem; and for the lovers of the beautiful in quiet out door and out-of-town life, this beautifully bound and attractively printed little volume will prove a companion and friend."—*Rochester Union and Advertiser.*

A NATURALIST'S RAMBLES ABOUT HOME. By CHARLES C. ABBOTT. 12mo. Cloth, $1.50.

"The home about which Dr. Abbott rambles is clearly the haunt of fowl and fish, of animal and insect life; and it is of the habits and nature of these that he discourses pleasantly in this book. Summer and winter, morning and evening, he has been in the open air all the time on the alert for some new revelation of instinct, or feeling, or character on the part of his neighbor creatures. Most that he sees and hears he reports agreeably to us, as it was no doubt delightful to himself. Books like this, which are free from all the technicalities of science, but yet lack little that has scientific value, are well suited to the reading of the young. Their atmosphere is a healthy one for boys in particular to breathe."—*Boston Transcript.*

DAYS OUT OF DOORS. By CHARLES C. ABBOTT. 12mo. Cloth, $1.50.

"'Days out of Doors' is a series of sketches of animal life by Charles C. Abbott, a naturalist whose graceful writings have entertained and instructed the public before now. The essays and narratives in this book are grouped in twelve chapters, named after the months of the year. Under 'January' the author talks of squirrels, muskrats, water-snakes, and the predatory animals that withstand the rigor of winter; under 'February' of frogs and herons, crows and blackbirds; under 'March' of gulls and fishes and foxy sparrows; and so on appropriately, instructively, and divertingly through the whole twelve "—*New York Sun.*

THE PLAYTIME NATURALIST. By Dr. J. E. TAYLOR, F. L. S., editor of "Science Gossip." With 366 Illustrations. 12mo. Cloth, $1.50.

"The work contains abundant evidence of the author's knowledge and enthusiasm, and any boy who may read it carefully is sure to find something to attract him. The style is clear and lively, and there are many good illustrations."—*Nature.*

THE ORIGIN OF FLORAL STRUCTURES through Insects and other Agencies. By the Rev. GEORGE HENSLOW, Professor of Botany, Queen's College. With numerous Illustrations. 12mo. Cloth, $1.75.

"Much has been written on the structure of flowers, and it might seem almost superfluous to attempt to say anything more on the subject, but it is only within the last few years that a new literature has sprung up, in which the authors have described their observations and given their interpretations of the uses of floral mechanisms, more especially in connection with the processes of fertilization."—*From Introduction.*

New York: D. APPLETON & CO., 1, 3, & 5 Bond Street.

D. APPLETON & CO.'S PUBLICATIONS.

THE GARDEN'S STORY; or, *Pleasures and Trials of an Amateur Gardener.* By GEORGE H. ELLWANGER. With Head and Tail Pieces by Rhead. 12mo. Cloth, extra, $1.50.

"Mr. Ellwanger's instinct rarely errs in matters of taste. He writes out of the fullness of experimental knowledge, but his knowledge differs from that of many a trained cultivator in that his skill in garden practice is guided by a refined æsthetic sensibility, and his appreciation of what is beautiful in nature is healthy, hearty, and catholic. His record of the garden year, as we have said, begins with the earliest violet, and it follows the season through until the witch-hazel is blossoming on the border of the wintry woods. . . . This little book can not fail to give pleasure to all who take a genuine interest in rural life."—*New York Tribune.*

THE ORIGIN OF CULTIVATED PLANTS. By ALPHONSE DE CANDOLLE. 12mo. Cloth, $2.00.

"Though a fact familiar to botanists, it is not generally known how great is the uncertainty as to the origin of many of the most important cultivated plants. . . . In endeavoring to unravel the matter, a knowledge of botany, of geography, of geology, of history, and of philosophy is required. By a combination of testimony derived from these sources M. de Candolle has been enabled to determine the botanical origin and geographical source of the large proportion of species he deals with."—*The Athenæum.*

THE FOLK-LORE OF PLANTS. By T. F. THISELTON DYER, M.A. 12mo. Cloth, $1.50.

"A handsome and deeply interesting volume. . . . In all respects the book is excellent. Its arrangement is simple and intelligible, its style bright and alluring. . . . To all who seek an introduction to one of the most attractive branches of folk-lore, this delightful volume may be warmly commended.—*Notes and Queries.*

FLOWERS AND THEIR PEDIGREES. By GRANT ALLEN, author of "Vignettes of Nature," etc. Illustrated. 12mo. Cloth, $1.50.

"No writer treats scientific subjects with so much ease and charm of style as Mr. Grant Allen. The study is a delightful one, and the book is fascinating to any one who has either love for flowers or curiosity about them."—*Hartford Courant.*

"Any one with even a smattering of botanical knowledge, and with either a heart or mind, must be charmed with this collection of essays."—*Chicago Evening Journal.*

THE GEOLOGICAL HISTORY OF PLANTS. By Sir J. WILLIAM DAWSON, F.R.S. Illustrated. 12mo. Cloth, $1.75.

"The object of this work is to give, in a connected form, a summary of the development of the vegetable kingdom in geological time. To the geologist and botanist the subject is one of importance with reference to their special pursuits, but one which it has not been easy to find any convenient manual of information. It is hoped that its treatment in the present volume will also be found sufficiently simple and popular to be attractive to the general reader."—*From the Preface.*

New York: D. APPLETON & CO., 1, 3, & 5 Bond Street.

D. APPLETON & CO.'S PUBLICATIONS.

THE DOG IN HEALTH AND IN DISEASE. By WESLEY MILLS, M. D., D. V. S., author of "A Text-Book of Animal Physiology," "A Text-Book of Comparative Physiology," etc. With colored plate, 38 full-page cuts, and numerous other Illustrations. 12mo. Cloth, $2.25.

The author of this work has undertaken, in a clear, concise, untechnical way, to supply the large class of intelligent dog owners and breeders, and veterinarians, with the information necessary for the proper care, management, and treatment of the dog. His well-known reputation as a writer and lecturer on human and veterinary physiology, his special study of canine diseases, and his long experience as a breeder of dogs, insure a thorough and correct handling of the subject.

"The library of every one interested in the dog should contain a copy of this work." —*American Stock-keeper, Boston.*

"The numerous illustrations in the book have been drawn from various sources, especial pains having been taken to furnish models for judging the various breeds of dogs in the full-page plates."—*Montreal Witness.*

"Altogether the work is one of great use to the breeder and the veterinary student, and one that should find a place in every dog-man's kennel-shelf and library."—*Forest and Stream.*

"Very interesting and valuable."—*New York Times.*

"A practical protest against the treatment of dogs according to the light of the horse-doctor. The book is intended for all persons who breed, keep, or in any way take a special interest in the dog. . . . One half the book is devoted to the diseases of the dog. The symptoms and treatment are carefully given, and there is added a table of doses of the drugs found most efficacious. The volume is one to be cordially recommended."—*Philadelphia Inquirer.*

THE COMPARATIVE ANATOMY OF THE DOMESTICATED ANIMALS. By A. CHAUVEAU, M. D., LL. D., Inspector General of Veterinary Schools in France. Second English edition, translated and edited by GEORGE FLEMING, C. B., F. R. C. V. S., etc., Examiner in Anatomy for the Royal College of Veterinary Surgeons, etc. With 585 Illustrations. 1084 pages. 8vo. Cloth, $7.00.

"This work has long since earned for itself the position of foremost rank in its particular field, and, indeed, stands without a rival in completeness, scientific arrangement, accurate detail, and practical adaptability to the necessities of the veterinary and to the student of general anatomy. It is profusely illustrated and handsomely printed." —*Medical Record.*

"'Chauveau's Anatomy,' as it now stands, is easily first, as it is, in fact, indispensable. The reputation of its author and editor is sufficient to prove that this work is all that it claims to be, and it may now continue to be accepted as in every way worthy of the position accorded to it as the best on the subject."—*Therapeutic Gazette.*

"Invaluable to all who desire a scientific knowledge of the animals utilized by man in the performance of his work. All intelligent persons interested in animals will greatly enjoy the study of the volume. As already intimated, it is issued in the elegant style characteristic of Appletons' publications"—*American Lancet.*

New York: D. APPLETON & CO., 1, 3, & 5 Bond Street.

D. APPLETON & CO.'S PUBLICATIONS.

WORKS BY ARABELLA B. BUCKLEY (MRS. FISHER).

THE FAIRY-LAND OF SCIENCE. With 74 Illustrations. Cloth, gilt, $1.50.

"Deserves to take a permanent place in the literature of youth."—*London Times.*

"So interesting that, having once opened the book, we do not know how to leave off reading."—*Saturday Review.*

THROUGH MAGIC GLASSES and other Lectures. A Sequel to "The Fairy-Land of Science." Cloth, $1.50.

CONTENTS.

THE MAGICIAN'S CHAMBER BY MOONLIGHT.
MAGIC GLASSES AND HOW TO USE THEM.
FAIRY RINGS AND HOW THEY ARE MADE.
THE LIFE-HISTORY OF LICHENS AND MOSSES.
THE HISTORY OF A LAVA-STREAM.
AN HOUR WITH THE SUN.
AN EVENING WITH THE STARS.
LITTLE BEINGS FROM A MINIATURE OCEAN.
THE DARTMOOR PONIES.
THE MAGICIAN'S DREAM OF ANCIENT DAYS.

LIFE AND HER CHILDREN: Glimpses of Animal Life from the Amœba to the Insects. With over 100 Illustrations. Cloth, gilt, $1.50.

"The work forms a charming introduction to the study of zoölogy—the science of living things—which, we trust, will find its way into many hands."—*Nature.*

WINNERS IN LIFE'S RACE; or, The Great Backboned Family. With numerous Illustrations. Cloth, gilt, $1.50.

"We can conceive no better gift-book than this volume. Miss Buckley has spared no pains to incorporate in her book the latest results of scientific research. The illustrations in the book deserve the highest praise—they are numerous, accurate, and striking."—*Spectator.*

A SHORT HISTORY OF NATURAL SCIENCE; and of the Progress of Discovery from the Time of the Greeks to the Present Time. New edition, revised and rearranged. With 77 Illustrations. Cloth, $2.00.

"The work, though mainly intended for children and young persons, may be most advantageously read by many persons of riper age, and may serve to implant in their minds a fuller and clearer conception of 'the promises, the achievements, and claims of science.'"—*Journal of Science.*

MORAL TEACHINGS OF SCIENCE. Cloth, 75 cents.

"The book is intended for readers who would not take up an elaborate philosophical work—those who, feeling puzzled and adrift in the present chaos of opinion, may welcome even a partial solution, from a scientific point of view, of the difficulties which oppress their minds."—*From the Preface.*

New York: D. APPLETON & CO., 1, 3, & 5 Bond Street.

D. APPLETON & CO.'S PUBLICATIONS.

THE ICE AGE IN NORTH AMERICA, and its Bearings upon the Antiquity of Man. By G. FREDERICK WRIGHT, D. D., LL. D., F. G. S. A., Professor in Oberlin Theological Seminary; Assistant on the United States Geological Survey. With an appendix on "The Probable Cause of Glaciation," by WARREN UPHAM, F. G. S. A., Assistant on the Geological Surveys of New Hampshire, Minnesota, and the United States. New and enlarged edition. With 150 Maps and Illustrations. 8vo, 625 pages, and Index. Cloth, $5.00.

"Not a novel in all the list of this year's publications has in it any pages of more thrilling interest than can be found in this book by Professor Wright. There is nothing pedantic in the narrative, and the most serious themes and startling discoveries are treated with such charming naturalness and simplicity that boys and girls, as well as their seniors, will be attracted to the story, and find it difficult to lay it aside."—*New York Journal of Commerce.*

"One of the most absorbing and interesting of all the recent issues in the department of popular science."—*Chicago Herald.*

"Though his subject is a very deep one, his style is so very unaffected and perspicuous that even the unscientific reader can peruse it with intelligence and profit. In reading such a book we are led almost to wonder that so much that is scientific can be put in language so comparatively simple."—*New York Observer.*

"The author has seen with his own eyes the most important phenomena of the Ice age on this continent from Maine to Alaska. In the work itself, elementary description is combined with a broad, scientific, and philosophic method, without abandoning for a moment the purely scientific character. Professor Wright has contrived to give the whole a philosophical direction which lends interest and inspiration to it, and which in the chapters on Man and the Glacial Period rises to something like dramatic intensity."—*The Independent.*

". . . To the great advance that has been made in late years in the accuracy and cheapness of processes of photographic reproduction is due a further signal advantage that Dr. Wright's work possesses over his predecessors'. He has thus been able to illustrate most of the natural phenomena to which he refers by views taken in the field, many of which have been generously loaned by the United States Geological Survey, in some cases from unpublished material; and he has admirably supplemented them by numerous maps and diagrams."—*The Nation.*

MAN AND THE GLACIAL PERIOD. By G. FREDERICK WRIGHT, D. D., LL. D., author of "The Ice Age in North America," "Logic of Christian Evidences," etc. No. 69, International Scientific Series. With numerous Illustrations. 12mo. Cloth, $1.75.

Every one is interested in ancestry, yet the roots of family trees have not struck down to the Glacial period, and we are left to wonder regarding the manners and customs of our ancestors in the remote age of ice. Who and what these ancestors were, is told us in simple, entertaining, popular style by Professor Wright, and his fascinating narrative is re-enforced by a multitude of illustrations.

New York: D. APPLETON & CO., 1, 3, & 5 Bond Street.

D. APPLETON & CO.'S PUBLICATIONS.

ESSAYS UPON SOME CONTROVERTED QUESTIONS. By THOMAS H. HUXLEY, F. R. S., author of "Man's Place in Nature," etc. 12mo. Cloth, $2.00.

"Professor Huxley is one of the most vigorous and uncompromising polemics of the age. He is admittedly also one of the very ablest. In that debatable land which touches the confines of science on the one hand and the confines of religion on the other, there is no living man who can be called his superior. Professor Huxley takes no mysterious or incomprehensible position. He plants himself on facts. There is, he contends, no other safe or solid position. . . . He has compelled attention to the truths of science. He has made religion more intelligent. He has helped us to see that science and religion are not mutually destructive—that the God of Nature and the God of the Bible are one God."—*Christian at Work*.

' . . . Mr. Huxley's literary style, also, is singularly lucid, polished, graceful, and strong. There is no living writer of English more 'cunning of fence' in dialectics; none who has a better gift for clothing his ideas in perspicuous and elegant language. He may be an agnostic or an infidel, with all that is implied in the words; but his candid declarations are always uttered with the refinement of a gentleman and the clearness of a thorough scholar."—*Philadelphia Bulletin*.

THE NATURALIST IN LA PLATA. By C. H. HUDSON, C. M. Z. S., joint author of "Argentine Ornithology." With 27 Illustrations. 8vo, 388 pages. Cloth, $4.00.

"Mr. Hudson is not only a clever naturalist, but he possesses the rare gift of interesting his readers in whatever attracts him, and of being dissatisfied with mere observation unless it enables him to philosophize as well. With his lucid accounts of bird, beast, and insect, no one will fail to be delighted."—*London Academy*.

"Mr. Hudson is a scientist, it is true, but apparently not enough of a scientist to have lost the faculty of wonder and the knack of making readers wonder with him. He has worked on an ornithology of the Argentine States, but his case appears to be that of Audubon, and the results of avoiding the drudgery of science the same. At least, he has so many new and interesting and piquant things to tell about the insects, birds, and beasts found on the pampas, that few books of the sort, if any, which have been published since the work of Belt appeared, can vie with it in charm."—*New York Times*.

NEW EDITION OF

FRAGMENTS OF SCIENCE. By JOHN TYNDALL, F. R. S., author of "Sound," "Heat as a Mode of Motion," etc. Revised and enlarged edition. 2 vols. 12mo. Cloth, $4.00.

The first edition of Professor Tyndall's "Fragments of Science" was published some twenty years ago as a single volume, which was made up of a score or more of his detached essays, addresses, and reviews. The book was afterward revised, some of the papers recast, and from time to time new ones added until, the size of the work becoming somewhat unwieldy, the present two-volume edition was decided upon. This contains fifteen additional papers, and represents the author's latest changes and revisions. The volumes are uniform with "New Fragments," recently issued, and the three together include all the occasional writings which their author has decided to preserve in permanent form.

New York. D. APPLETON & CO., 1, 3, & 5 Bond Street.

D. APPLETON & CO.'S PUBLICATIONS.

EVOLUTION IN SCIENCE, PHILOSOPHY, AND ART.

A Series of Seventeen Lectures and Discussions before the Brooklyn Ethical Association. With 3 Portraits. 466 pages. 12mo. Cloth, $2.00. Separate Lectures, in pamphlet form, 10 cents each.

These popular essays, by some of the ablest exponents of the doctrine of evolution in this country, will be read with pleasure and profit by all lovers of good literature and suggestive thought. The principle of evolution, being universal, admits of a great diversity of applications and illustrations; some of those appearing in the present volume are distinctively fresh and new.

CONTENTS.

1. *Alfred Russel Wallace* . . . . . . By EDWARD D. COPE, Ph. D.
2. *Ernst Haeckel* . . . . By THADDEUS B. WAKEMAN.
3. *The Scientific Method* . . . . . . By FRANCIS E. ABBOT, Ph. D.
4. *Herbert Spencer's Synthetic Philosophy.* By BENJ. F. UNDERWOOD.
5. *Evolution of Chemistry* . . . . . . By ROBERT G. ECCLES, M. D.
6. *Evolution of Electric and Magnetic Physics.*
   By ARTHUR E. KENNELLY.
7. *Evolution of Botany* . . . . By FRED J. WULLING, Ph. G.
8. *Zoölogy as related to Evolution* . . . By Rev. JOHN C. KIMBALL.
9. *Form and Color in Nature* . . . . By WILLIAM POTTS.
10. *Optics as related to Evolution* . . . By L. A. W. ALLEMAN, M. D.
11. *Evolution of Art* . . . . By JOHN A. TAYLOR.
12. *Evolution of Architecture* . . . . . By Rev. JOHN W. CHADWICK.
13. *Evolution of Sculpture* . . . . . . By Prof. THOMAS DAVIDSON.
14. *Evolution of Painting* . . . . . . By FORREST P. RUNDELL.
15. *Evolution of Music* . . . . By Z. SIDNEY SAMPSON.
16. *Life as a Fine Art* . . . . By LEWIS G. JANES, M. D.
17. *The Doctrine of Evolution: its Scope and Influence.*
    By Prof. JOHN FISKE.

"A valuable series."—*Chicago Evening Journal.*

"The addresses include some of the most important presentations and epitomes published in America. They are all upon important subjects, are prepared with great care, and are delivered for the most part by highly eminent authorities"—*Public Opinion.*

"As a popular exposition of the latest phases of evolution this series is thorough and authoritative."—*Cincinnati Times-Star.*

New York: D. APPLETON & CO., 1, 3, & 5 Bond Street.

D. APPLETON & CO.'S PUBLICATIONS.

NEW FRAGMENTS. By JOHN TYNDALL, F. R. S., author of "Fragments of Science," "Heat as a Mode of Motion," etc. 12mo. 500 pages. Cloth, $2.00.

Among the subjects treated in this volume are "The Sabbath," "Life in the Alps," "The Rainbow and its Congeners," "Common Water," and "Atoms, Molecules, and Ether-Waves." In addition to the popular treatment of scientific themes, the author devotes several chapters to biographical sketches of the utmost interest, including studies of Count Rumford and Thomas Young, and chapters on "Louis Pasteur, his Life and Labors," and "Personal Recollections of Thomas Carlyle."

"Tyndall is the happiest combination of the lover of nature and the lover of science, and these fragments are admirable examples of his delightful style, and proofs of his comprehensive intellect"—*Philadelphia Evening Bulletin.*

"The name of this illustrious scientist and *littérateur* is known wherever the English language is the mother tongue, or is even freely spoken. Whatever he does or says comes with a stamp of authority as from one who speaks with power, knowing whereof he affirms. He is able and effective, both as a talker and writer, as scientist or teacher. To those who know anything of Prof. Tyndall's life and labors, scientific or literary, it is superfluous to say that his utterances bring his hearers or readers face to face with the latest knowledge on the subject he discusses."—*New York Commercial Advertiser.*

MORAL TEACHINGS OF SCIENCE. By ARABELLA B. BUCKLEY, author of "The Fairy-Land of Science," "Life and her Children," etc. 12mo. Cloth, 75 cents.

"The book is intended for readers who would not take up an elaborate philosophical work—those who, feeling puzzled and adrift in the present chaos of opinion, may welcome even a partial solution, from a scientific point of view, of the difficulties which oppress their minds."—*From the Preface.*

MAX MÜLLER AND THE SCIENCE OF LANGUAGE. A Criticism. By WILLIAM DWIGHT WHITNEY, Professor in Yale University. 12mo. 79 pages. Paper cover, 50 cents.

This critique relates to the new edition of Prof. Müller's well-known work on Language. "For many," says Prof. Whitney, in his preface, "the book has been their first introduction to linguistic study ; and doubtless to a large proportion of English-speaking readers, especially, it is still the principal and most authoritative text-book of that study, as regards both methods and results. A work holding such a position calls for careful criticism, that it may not be trusted where it is untrustworthy, and so do harm to the science which it was intended to help."

"This caustic review of Max Müller's latest edition of his 'Science of Language' will command attention for more and higher merits than its brilliant criticism. It upholds a theory of language and of its development which, though not taught by Max Müller, is held by the great masters of linguistic science. The reader not versed in the science, nor well read in its controversial literature, will get from this *brochure* a conception of the critical points of the subject which he might miss in the reading of many larger and more systematic treatises."—*The Independent,* New York.

New York: D. APPLETON & CO., 1, 3, & 5 Bond Street.

D. APPLETON & CO.'S PUBLICATIONS.

MODERN SCIENCE SERIES.
Edited by Sir JOHN LUBBOCK, Bart., F. R S.

The works to be comprised in the "Modern Science Series" are primarily not for the student, nor for the young, but for the educated layman who needs to know the present state and result of scientific investigation, and who has neither time nor inclination to become a specialist on the subject which arouses his interest. Each book will be complete in itself, and, while thoroughly scientific in treatment, its subject will as far as possible be presented in language divested of needless technicalities. Illustrations will be given whenever needed by the text. The following are the volumes thus far issued. Others are in preparation.

THE CAUSE OF AN ICE AGE. By Sir ROBERT BALL, LL. D., F. R. S., Royal Astronomer of Ireland, author of "Starland." 12mo. Cloth, $1.00.

"Sir Robert Ball's book is, as a matter of course, admirably written. Though but a small one, it is a most important contribution to geology."—*London Saturday Review.*

"A fascinating subject, cleverly related and almost colloquially discussed."—*Philadelphia Public Ledger.*

THE HORSE: A Study in Natural History. By WILLIAM H. FLOWER, C. B., Director in the British Natural History Museum. With 27 Illustrations. 12mo. Cloth, $1.00.

"The author admits that there are 3,800 separate treatises on the horse already published, but he thinks that he can add something to the amount of useful information now before the public, and that something not heretofore written will be found in this book. The volume gives a large amount of information, both scientific and practical, on the noble animal of which it treats."—*New York Commercial Advertiser.*

THE OAK: A Study in Botany. By H. MARSHALL WARD, F. R. S. With 53 Illustrations. 12mo. Cloth, $1.00.

"An excellent volume for young persons with a taste for scientific studies, because it will lead them from the contemplation of superficial appearances and those generalities which are so misleading to the immature mind, to a consideration of the methods of systematic investigation."—*Boston Beacon.*

"From the acorn to the timber which has figured so gloriously in English ships and houses, the tree is fully described, and all its living and preserved beauties and virtues, in nature and in construction, are recounted and pictured."—*Brooklyn Eagle.*

ETHNOLOGY IN FOLKLORE. By GEORGE LAWRENCE GOMME, F. S. A., President of the Folklore Society, etc. 12mo. Cloth, $1.00.

This book is an attempt to ascertain and set forth the principles upon which folklore may be classified, in order to arrive at some of the results which should follow its study, giving the subject the importance it deserves in connection with researches in ethnology.

New York: D. APPLETON & CO., 1, 3, & 5 Bond Street.

www.ingramcontent.com/pod-product-compliance
Lightning Source LLC
Chambersburg PA
CBHW031848220426
43663CB00006B/535